For Jon,

With imm

lifetime of

mentorship, inspired by your sense of decency, your values and your ability to speak truth to power. I owe a lot of my ability to think, to analyze and to process to you — and for that, and for so many other things — I'll always be grateful.

میں آپ کی محبت، دوستی، سرپرستی اور تعاون کا تہہ دل سے ممنون ہوں اور عمر بھر اس محبت اور علم دوستی کو یاد رکھوں گا۔

احسان مند

محمد حامد زمان

Muhammad H. Zaman

Geneva

June 27th, 2018.

BITTER PILLS

BITTER PILLS

The Global War on Counterfeit Drugs

Muhammad H. Zaman

Oxford University Press is a department of the University of Oxford. It furthers the University's objective of excellence in research, scholarship, and education by publishing worldwide. Oxford is a registered trade mark of Oxford University Press in the UK and certain other countries.

Published in the United States of America by Oxford University Press
198 Madison Avenue, New York, NY 10016, United States of America.

Library of Congress Cataloging-in-Publication Data
Names: Zaman, Muhammad H. (Muhammad Hamid), author.
Title: Bitter pills : the global war on counterfeit drugs / Muhammad H. Zaman.
Description: New York, NY : Oxford University Press, 2018. |
Includes bibliographical references and index.
Identifiers: LCCN 2017046698 | ISBN 9780190219444
Subjects: | MESH: Counterfeit Drugs | Fraud—prevention & control |
Drug Contamination—prevention & control |
Pharmaceutical Preparations—standards | Quality Control
Classification: LCC RS189 | NLM QV 773 | DDC 338.4/76153—dc23
LC record available at https://lccn.loc.gov/2017046698

9 8 7 6 5 4 3 2 1

Printed by Sheridan Books, Inc., United States of America

To Afreen, Rahem, and Samah

Author's Note

While from a technical standpoint, counterfeit and substandard refer to different problems, in many sectors of media and popular press, the term counterfeit is used as an all encompassing term, and is most commonly used to describe the problem of bad drugs. The title of the book uses the term counterfeit as a reflection of the general discourse on the problem.

Contents

Acknowledgments

This book has many stories. Some I experienced firsthand in back alleys in Africa and in the boardrooms in Europe, and some were revealed to me by manuscripts from the Middle Ages. Behind each tale, and my experience in understanding that story, are people whom I am deeply indebted to. These people have made it possible for me to understand both the context and the complexity of the problem, and to share my observations through this book.

The person who has taught me the most during the book-writing process is Barbara Moran. Through her characteristic wit and deep insights, she taught me how to tell a story, how to create flow in a science book for a broad audience, and how to keep the themes alive, even when the technical content gets heavy. She also helped me connect with my wonderful agent, Michelle Tessler, who has been incredibly patient with me, yet persistent enough to see that I finish the book. Another friend, Sara Rimer, has been a champion and supporter of this project and has always been there for candid advice in all of my science communication projects. I am also grateful to my editorial team, Jeremy Lewis and Anna Langley at Oxford University Press, for helping me make my arguments sharper and my text more accessible. My editor at Project Syndicate, Joanna Rose, has helped me improve my writing tremendously, and has given me the opportunity to share my perspective with a broad global audience, and I am grateful to her for her support and mentorship.

Over the course of the last few years, there have been many mentors who have been instrumental in shaping both this book and the person I am. Among them, Anthony Boni at USAID has been the most influential person in helping me understand the field of global development, particularly with respect to pharmaceutical quality. I am deeply indebted to his wisdom and mentorship through the years. Dinners and receptions at his place will always have a special place in my heart. Colleagues at United States Pharmacopeia, including Patrick Lukulay and Kennedy Chibwe, helped me understand the challenges of drug quality in Africa. Jude Nwokike has always been there for a quick chat, or a long meeting, to help me understand where the challenges and opportunities lie. Experts at USAID's Center for Accelerating Innovation and Impact, Wendy Taylor, Marissa Leffler, Karen Clune, Emily Hillman, and Vinesh Kapil, have been champions of my work and also the most objective critics I could find. Other mentors who have shaped my thinking in this area include Peter Singer, Haitham El-Noush, and Frederik Kristensen. Their inspiration has helped me understand the need for why innovation in improving drug quality is needed and why things are so hard to scale. Despite the challenges, their motivation and passion are making the world a better place. Paul Newton is one of the world leaders in the field of drug quality, and I am deeply grateful to him for sharing with me papers and resources that I had not seen before.

I could not have completed this book without the support, patience, time, and incredible help from in-country partners. From Pakistan to Peru, India to Indonesia, Senegal to Switzerland, many outstanding researchers, pharmacists, public health professionals, and public officials provided input on my work and were incredibly generous with their time. Khalid Saeed Bukhari in Pakistan stands out in particular for helping me understand the complex dimensions of drug regulation in the country. Colleagues at relevant government agencies in Kenya, Tanzania, Ghana, and Indonesia helped me to understand the various dimensions of procurement, supply, and demand for essential medicines. They were also candid about what ails the system and how to change the status quo. Some of these individuals, given the sensitive nature of the work that they do, and due to potential consequences of their candid opinions, requested that I maintain their anonymity. I will honor their wishes but will always be indebted to them for their support.

Leaders in pharmaceutical industry, in both research-based multinationals and generic industry, took time out to speak to me multiple times. I am grateful to them for sharing their perspective.

Howard Hughes Medical Institute is a special place with very special people. Sarah Simmons, David Asai and Sean Carroll have been supportive of all of my adventures, and they gave me both a sense of purpose in my pursuits to understand some of the most pressing issues of our time and the support to bring these challenges to the classroom. Boston University, and my colleagues both in Engineering and in Public Health, have been incredibly supportive of all of my efforts. Jim Collins has been a mentor and an incredible supporter of my work. I have learned so much from him. Jonathan Simon, Christopher Gill, Don Thea, and Sandro Galea at the Boston University School of Public Health have always been there to help me navigate the complexity of global public health. Veronika Wirtz and Richard Laing taught me everything I know about how access to medicines work in the low-income countries. My current and former department chairs, John White and Solomon Eisenberg, my Dean Kenneth

Lutchen, as well as the provost Jean Morrison and the university president, Robert Brown, have given me the freedom and resources to pursue projects that often do not fall within the scope of traditional engineering. This freedom and support have enabled me to experience firsthand the global challenges that I talk about in this book. I am honored by their support and mentorship over the years.

This book, or any part of my research, would not be possible without the talent, hard work, and creativity of my research group. I have been exceptionally fortunate to work with a team of students and scholars who inspire me every day and have made it possible for me to ask interesting questions and to seek their answers. Darash Desai, Katie Clifford, Zohar Weinstein, Wolfgang Krull, Andrea Fernandes, Nga Ho, Grace Wu, Andrew Acevedo, and Atena Shemirani have helped me every step of the way in understanding the stubborn challenge of poor-quality medicines around the world. Katie, in addition to being an outstanding member of the team, also took time out to help me with research on PQM history, read the manuscript that helped me clarify numerous confusing parts.

My team at home has been just as instrumental in making this book possible. My wife, Afreen, is not just the most loving partner anyone could imagine, she is also my best friend and an unending source of my inspiration. She has been incredibly patient, supportive, and understanding as I have had to go to far-off places for my work. Our son, Rahem, and our daughter, Samah, have always provided me with the reason to take an earlier flight back. Being home with them is the greatest pleasure of my life.

This book ultimately is about hope for a better and more equitable world. No one provides me with that hope more than my family. It is to them, with all my love, that I dedicate this book.

Prologue

I never knew what the D in D. Watson Chemist stood for.

The D. Watson Chemist and Drugstore was a store in a bigger market, called the Super Market, in the heart of Islamabad. Islamabad in the 1980s was the sleepy capital of Pakistan and also was my hometown. Was it related to Doctor Watson, a trusted friend and the narrator of Mr. Holmes's achievements in Sir Arthur Conan Doyle's classics? Or was it referring to something else? Some said that it was part of a foreign chain, which seemed unlikely because there were very few foreign chain stores in Pakistan, and almost none in Islamabad. Some felt that the store was owned by some local Christian family, but the Christian community in Pakistan is not particularly well off, and most do not have a foreign-sounding last name. Watson would not be a Christian name in Pakistan. Ultimately, I gave up and never figured out what the D was for, or even who Mr. or Ms. Watson was, and what was the relationship of this name with a store that was operated and staffed by Pakistanis.

But D. Watson was a very special place. In a town of about two hundred thousand people, and one that was known to fall asleep by 9 PM, D. Watson was open until late. It had bright white lights that lit up the whole store. The staff was efficient, knowledgeable, and courteous. This is where one would find the elite of Islamabad fill their prescriptions. This was also the place where the foreigners, who were part of the diplomatic mission, or were part of various NGOs, purchased their medicines. The store had drug brands that were familiar to the foreign clientele. It also employed staff that could make do in English.

To the Islamabad residents of the 1980s, D. Watson represented everything that was right with the West in our minds. It was polished, well stocked, well staffed, and very clean. The foreign name gave it an aura of mystery and authority.

More than anything it had one priceless commodity that it cashed in everyday: trust.

The house that I grew up in in Islamabad was not particularly close to D. Watson. Yet we went there more than we went to the pharmacy that was a lot closer to our house. The drive to D. Watson was a bit of a ride but that didn't matter when it came to getting important prescriptions filled. We would get the aspirins and the cough syrups at the store that was closer, but if it was anything serious, it had to be at the D. Watson Chemist and Drugstore in the Super Market. The vaccines, if needed, were almost exclusively bought at D. Watson. There was no question of going anywhere else. We weren't among the elite of the city, but the brand mattered to my mother.

I grew up with that assumption, that some chemist stores are better, more reliable, and more trustworthy than others. Some sell good products, others don't, and they all operate in the same town. In my mind, this was the natural order of things, a hierarchy of quality that allowed coexistence. For me, it represented an equilibrium between the good and the bad, and a marketplace of health commodities with a spectrum of quality.

It was much later in life, after I had moved to the United States that I realized that it was not supposed to be this way. I remember when I asked a friend which pharmacy was better in terms of its products, he had no idea what I meant. I tried multiple times, and then he and I both gave up. Whether I was in Athens, West Virginia, or Russellville, Arkansas, or Chicago or Boston, the Tylenol was the same and so was the quality of prescription medicines. The pharmacy in the inner city had the same quality of drugs as the one in the most exclusive neighborhoods.

But this was not how it was supposed to work. There was no spectrum of quality here. Something did not quite add up.

This book is an effort to find answers to the questions I have had for so long, about trust, technology, and why quality is so elusive.

1

Cities in Crisis

My cell phone rang on the morning of January 13, 2015. On the other end was a friend, who had worked in the Non-Governmental Organization (NGO) sector and who was aware of my research work in Ghana on testing substandard and counterfeit medicines. With a sense of panic in his voice, he asked me if I had read the news about the central medical stores in Tema, near Accra? I hadn't. He told me that the entire facility, with all its supplies and medicines, had burned down, the cause unknown.[1] He suspected arson. My heart sank.

Tema is only 25 kilometers from Accra, but the conditions of the roads make it a much longer journey. The poisonous mix of traffic, accidents, and road conditions made it a half-day journey when I visited the place in March 2014. During our trip, we were greeted warmly by the manager of the facility, who had shown us stacks upon stacks of medicines and other medical supplies. The supplies ranged from boxes of medicines to patient beds, vaccines to baby formula. There were even new shiny Japanese cars parked in the parking lot, ready to be used for various Ministry of Health services. A significant number of medical supplies were donated and had colorful stamps and logos in foreign languages showing the origin of the donating agencies and their country affiliations. Some materials were also purchased by the government and had reached the facility at Tema through various customs points.

Beyond the parking lot, with the shiny vehicles, were a set of large warehouses, which served as the storage facility for medical supplies. Our host, the facility manager, walked with us, found a key that matched the lock, and opened the warehouse for us. The warehouse was a large structure with high ceilings. From the outside it looked like an impressive facility, but as we stepped in, it painted a different picture. This brick-and-mortar structure was in a state of disrepair. There was no power, and we had to rely on a few rays of sunlight to find our way between the mazes of boxes arranged in rows. Light was the least of the problems we had, compared to what we were about to experience. The much bigger problem was the stifling heat, which had amplified due to lack of ventilation. I looked at the temperature app of my phone to see that the temperature was reaching 95°F. There were lots of challenges at the facility, and the manager of the facility wanted us to look at them firsthand, so we could understand his frustration and the challenges he faced on a daily basis. This was my only trip to the central medical stores in Ghana.

While central medical stores are found in much of sub-Saharan Africa, they are a remnant of the colonial era. The original idea was to have central facilities and storage sites for maintaining control and distribution in a highly centralized system of government. These were also the points that coordinated arrival of shipments from the colonial powers in Europe. For the colonists, it was much easier to send (or sell) supplies to a central point than deal with the complexity of local markets.

Most central medical stores in Africa date back to the early or mid-twentieth century, before the countries gained independence. In Kenya, for example, they date back nearly a hundred years, when they were established in 1915 as headquarter medical stores. Going through various stages of evolution and incarnation, the term "central medical store" in Kenya was christened as such in 1970.[2] Today, they are called the Kenya "medical supplies authority." Similarly in Mauritius they date back to 1945, before the independence of the country.[3] In Sudan, they date back to 1935.[4] Most have not only retained their form in one way or another, but also have grown over time. Like in Tema, most central medical stores are now located outside the city, in industrial areas.

Central medical stores serve as the nerve centers and main storage facilities in many African countries, where drugs are stored and logged before they reach the markets or provinces. After arrival at the dock, the drugs arrive at the medical store and while some are sent for testing to national labs, most are not. They are also sites for implementation of various disease-management programs funded by international donors. The drugs and medical supplies, through an internal demand and procurement process, reach various provinces and the process continues down to the district level.

Ghana has long been considered a model for peace, harmony, and stability in West Africa. It is a stable country in a region that is marred by violence, civil war, long dictatorships, and, most recently, terrorism. Being an English-speaking country, Ghana was where my team and I have worked for the last several years. The central medical stores would have been the central location to pilot the technology we have been working on, to test drug quality, before the drugs make it to the market. This would have been the site to see if we could detect the quality of drugs before the public was put at risk.

Now it was all gone.

The loss of the central medical store in Ghana meant the loss of not only millions of dollars' worth of drugs, but also the main mechanism by which drugs enter the country, and are logged, albeit in a system that still relied on paper and old ledgers. In the absence of this store, there was no mechanism by which legally procured drugs could enter this stable country. There has been some progress in determining the cause of the fire, now widely believed to be arson, but many important questions remain about motivation, collusion, and lack of security.[5] This lack of transparency affects public trust in the system and leads to conspiracy theories and the spread of misinformation, issues that further hurt the ability of governments to safeguard drug quality and build confidence in its methods.

A similar call had interrupted my morning in January 2012. This time the story was about Lahore, an ancient city in the northeastern part of Pakistan. A city of art, learning, food, and music, Lahore is the cultural center of Pakistan.

A mystery illness had started to kill patients at a public hospital in the heart of Lahore. Patients at the Punjab Institute of Cardiology started dying toward the end of December and early January.[6] The deceased came from lower socioeconomic backgrounds, were largely dependent on the government healthcare program, and had been getting the said medicine from the hospital for quite a while. Families were shattered as their loved ones started dying mysteriously. Among them was Ashiq Hussain, a loving grandfather and a man known for his kind manners and a big heart, who died due to inexplicable causes. He was ill but no one expected him to die suddenly.[7]

Prone to conspiracy theories, the media first blamed it on the sinister elements in the society. Terrorism was also a possibility, given Pakistan's recent history and the lack of security available to the common folk. Some others suggested links to India, the archenemy to the east with whom Pakistan has had three large and countless small skirmishes. Some doctors suggested that this was arsenic or heavy-metal poisoning, since they suspected something was not quite right with the way patients were reacting to the treatment. The numbers continued to swell, reaching near two hundred, while the cause remained a mystery.

The pressure from the public and the media was intense. Buckling under this pressure and the public outcry, the government of the Punjab Province (Lahore being the capital of the Punjab Province) decided to send samples of the drugs that had been given to the patients to the United Kingdom for testing. The local drug-testing labs were not trusted and their results, some argued, could have been manipulated by the government.[8]

Mystery shrouded the entire process, with some experts questioning the governance of the province while others were unsure of why the samples were sent abroad.[9] Eventually it was determined that the cause of the adverse reaction was due to contamination of the antihypertensive medicine, Isotab, with an antimalarial medicine, pyrimethamine, leading to deposits of the antimalarial in the bone marrow of the patients.[10] The detailed reports, submitted to the courts, suggest that up to 14 percent of the antimalarial was added to the drug, a lethal amount given the patients' history and their ailments. The drug in question was manufactured by Efroze Pharmaceuticals, a reputed company based in the port city of Karachi, and a hub of local drug manufacturing.

During the early days of the crisis, the company executives were put on the exit control list, Pakistan's version of a "do not fly" list, to stop them from leaving the country. But, quietly, after a couple of weeks the names were cleared. The federal investigative agency had also blocked the company website, which also came back online not too long after the incident. The drug company is now fully operational, and no one quite seems to fully understand exactly what penalties have been imposed, or what happened to the ones that were being imposed. Just as with the initial scandal, the events post investigation are shrouded in mystery, misinformation, and confusion.

The exact cause of why the drug supply was contaminated is also unclear and points to a breakdown of internal checks and balances as well as a lack of seriousness in maintaining quality control on part of the manufacturer. To investigate the matter in detail, a World Health Organization (WHO) team, along with local WHO representatives, created a tribunal, talked to dozens of people, and analyzed the evidence. They also visited the facility in Karachi, looked at the existing safety checks, and also analyzed the quality assurance methods in the public sector.[10]

The report provided detailed evidence regarding how barrels of white powder of the two drugs, which looked similar, got mixed up. Some of the technicians suspected things were not quite right and while they did point that out and tried to raise an alarm, the upper management did not act appropriately. The problem at Efroze was not with intentional falsification or an intent to deceive the patient, but with a breakdown of checks and balances. This breakdown was not just at various points within the pharmaceutical company, in quality control, but also with the public authorities tasked with testing drugs. Despite major lapses in safety, quality assurance, and adherence to best practices, they had been given multiple clean slates during inspection, including some not too long before the incident.

The government of the Punjab Province and the federal government in Pakistan, in response to the crisis, or perhaps in response to the public outrage of the crisis, also tried to create various agencies and task forces,[11] but large-scale measures to improve quality are still pending.

A few months after the Punjab Institute of Cardiology crisis, another crisis erupted in the same city of Lahore and the nearby city of Gujranwala.[12] This time it was a cough syrup, Tyno, that was associated with the deaths of dozens of people. In addition to Lahore, people also died in Gujranwala. The exact number of those who die in a country like Pakistan can never be known because people bury their dead quickly and any culture to investigate the cause of death remains nonexistent.

This crisis was too close to the previous one, in time and geographic proximity, and put significant pressure on the government. In order to deflect attention and show that it was not the fault of the public health department, the government initially blamed those within the urban slums, who are among the poorest in the country and belong to minority Christian groups. The position of the government was that these slum dwellers were addicted, and because alcohol is not easily available, in order to satisfy their substance addiction they were overdosing on the said syrup. The slum dwellers were an easy target with their little political representation, poor image, and minimal support among the civil society in Pakistan. Blaming the victims, particularly the poorest of the poor, is not unheard of in Pakistan. This is particularly true for those who may also be indulging in practices, such as addiction or substance abuse, that are looked down upon in society. The official government line to blame the slum dwellers was quick, well before any investigation or postmortem was carried out.[13] Yet this time there was a quick reaction, particularly on social media and from the family of one of those who had died. The brother of the dead man challenged the government's narrative and his position went viral on social media.[14]

The government then changed its position and took a more ambiguous line, promising a serious investigation in the near future. The government hoped that with time and a new crisis emerging somewhere else in some other part of the country, people would forget. They were right.

After investigations, by both Pakistani and UK authorities, spurious ingredients were detected in the syrup. Though the test reports were never made public,[13] the government blamed various groups, from manufacturers to opposition for blowing the situation out of proportion and jumping to conclusions to

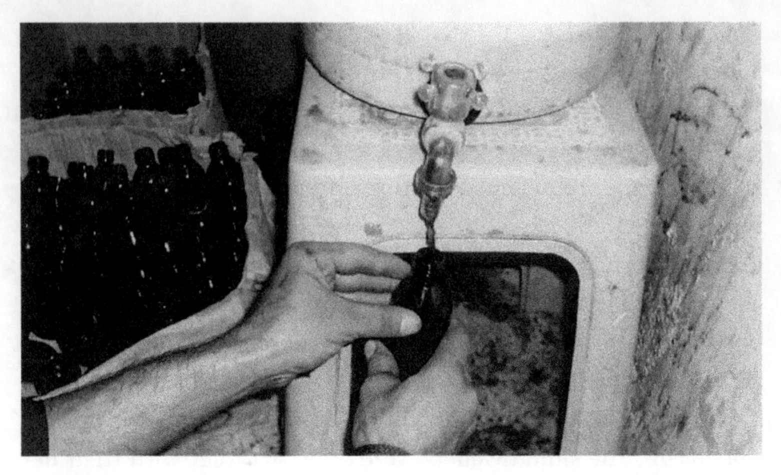

FIGURE 1.1.

Manufacturing at a cough syrup factory near Lahore, Pakistan. Lahore has been the site of several recent high-profile incidents due to poor-quality medicines. Reproduced with permission from Pfizer.

malign its good performance. Little has been done since to understand the cause of the problems.

The problem of counterfeit drugs is not just a problem of low-income countries and large urban cities in these nations. Cities, small and large, in more affluent countries are not immune either. Cape Cod, Massachusetts, swells with tourists during the summer and shrinks into itself during the winter. Most folks who live on the Cape year-round cobble together a few jobs to make a living. They rent houses and sell ice cream in July; they scrub boat hulls and paint bathrooms in January.

In this way, forty-five-year-old David Nailen was much like his neighbors in East Falmouth, except his side job was far more profitable than most of theirs. And it was illegal. Nailen ran a thriving business buying counterfeit Viagra from Hong Kong and selling it on Craigslist at a huge markup. In May 2013, after months of painstaking police work by US Customs officials, he was indicted for trafficking in counterfeit goods, and a warrant was issued for his arrest.[15] In July, a quick Google search revealed that Nailen was back in business. "Get Pfizer Viagra 100mg 4 pills $40," his Twitter account reads. "No generic rubbish. The original and the best."

An online search about counterfeit drugs or its adverse effects leads to dozens of stories throughout the world. There is a new story every week. Some are about the harmful, perhaps even deadly effects, and some stories are about shipments of counterfeit drugs with millions of packets of junk in them.

In June 2012, one of the largest shipments of counterfeit drugs was seized in Luanda, the bustling capital of Angola.[16] Nearly 1.4 million packets of substandard Coartem, a leading antimalarial manufactured by Novartis, were seized. To put the size in perspective, this shipment would have been enough to treat half of the annual cases of malaria in the whole country. The fakes were stored in a shipment of loudspeakers that originated in Guangzhou, and the shipment had reached through an intricate and opaque network of distributors and suppliers to Angola.

Angola is hardly an anomaly.[17] Neither is Africa alone in its challenges of shipments of bad drugs that reach its ports. Similar incidents of large shipments confiscated in a variety of countries, in nearly every continent, paint a global picture of complexity that is hard to track. The Angolan consumers interviewed by the *Wall Street Journal* said that there is a real demand for cheap knock-offs, as sometimes they do work. Some argued that the demand for substandards is in part driven by hospitals running out of stock. The knock-offs are a lot easier to access. While drug outages are usually denied by the authorities, they occur routinely due to the fragmented supply chains. In interviews with consumers and even pharmacists in multiple countries, from

Pakistan to Tanzania, Ghana to Nepal, there is never any denial of stock-outs, forcing patients to buy drugs from the private sector.[18] Many of the pharmacies, which are peppered around the public hospitals, often carry drugs that were meant to be free for the public but were never registered in the country in the first place. It is not exactly easy to know whether the stock-outs are real, and due to limited resources and poor management, or artificial, created due to corruption and collusion.

Scanning various news items, in print, electronic or social media, about substandard and counterfeit drugs, can be overwhelming. The incidents are often in different places and point to differing circumstances. Analyses of these stories in various cities point to a number of commonalities, but they also suggest the multifaceted nature of the problem that makes it hard to track. The cities mentioned above point to different dimensions of the problem.

The situation in Tema points to the challenge of security and safeguarding the supply of essential and lifesaving medicines. While that is not the only challenge, the fire in Tema points to both a failure of security and the vulnerability of the model of central medical stores. By creating a central repository, which functions as both the warehouse and the nerve center, the system is precariously balanced on a site that is vulnerable to external threats, terrorism, internal corruption, and other outside factors. It also points to how poor infrastructure affects the ability of the country to maintain a healthy supply chain and ensure its integrity. If, for example, there is complete breakdown of power, the entire stock in the central medical store becomes vulnerable. Vaccines and medicines that need continued cold storage may lose all of their potency, and with that the investment that has been made to procure these drugs also disappears. With little or no testing available, the same substandard and degraded drugs would enter the market far and wide in the country. While a centralized system may reflect a central donor-centric worldview that is easier to track and manage, it also reflects the vulnerability of the system that is highly dependent on a single site or a few key locations. The central medical stores are also not immune from theft.[19] There are also criticisms about having one central store, close to the capital, which, in the presence of a weak infrastructure and corruption, would mean that towns and villages far from the capital may not see drugs and supplies reach there in due time.[20]

The concept of central medical stores is not just about a warehouse that works in a vacuum or isolation. It is inherently connected to the rest of the public health system. The connections between the central medical store to the hospitals and regional medical stores and pharmacies are vital for timely availability and access of medicines. The connection mechanism (even if it is poorly structured) from the central medical store, on the one hand, feeds into the national labs, and on the other, connects to regional sites, hospitals, and pharmacies. In some cases, these medical stores also supply the private-sector pharmacies.

The central medical store fire in Tema points to a tough and difficult road ahead for Ghana, a path that is uncertain and is being debated in Ghanaian public health circles.[21] Creating a new central medical store certainly is a pricy endeavor, and the utility of all that investment is somewhat questionable. The investment, should it be made, would be in a system that is looking more toward the past than the future. The alternative is not obvious either. A possible alternative may be to have strong regional medical centers and equivalents of medical stores in various regions and provinces. This would require creating systems in other provinces that would have the capacity and capability to receive direct shipments, do some kind of testing, and then release the medicines to the local supply chains. Perhaps more efficient, this is also by no means a cheap alternative. This would also mean bringing some kind of regular testing services and higher-quality technical capacity to the regional sites that currently does not exist. Among the various options being debated, this is the road Ghana seems to be taking, though the exact roadmap is unclear. So is the timeline of the next stage. Perhaps a bigger question is what happens in the interim? Would this lead to drugs entering the market that go under the radar? Would the mechanisms that are created, in the interim, to bring medicines from abroad include sufficient testing? What about poor-quality drugs that enter the market, create new pathways for those with nefarious interests, and compromise public health and create long-term problems such as drug resistance? Would those pathways be entrenched in the system and stay like that even after the central medical stores have been re-created? The impact of the loss of the central medical stores is going to be far-reaching, and it will take both time and lots of effort before we fully understand the impact. As the model of central medical stores is repeated in a number of countries across the African continent and beyond, the problem in Tema reflects a broader challenge that can undermine the efforts of the government and international donors in maintaining and safeguarding the pharmaceutical supply lines.

The situation in Lahore with the incident at the Punjab Institute of Cardiology also points to a serious crisis, but one that is distinct from the problem in Tema. This crisis is also multidimensional in its origin and impact. First, it points to lack of robust regulatory oversight within the company and in the country. Regular testing within the company and in the public sector is seriously deficient. From various reports and tribunals, it is now clear that the drugs that had reached the hospital did not meet the quality standards that were expected of the company. Contamination of an antihypertensive with an antimalarial shows a deep and serious problem during manufacturing that went unchecked until it was too late.

Investigators found that the quality control and manufacturing staff at Efroze realized that something was amiss when their antimalarial had a lower-than-normal concentration. So not all of the antimalarial from the stock was entering the drug. This was due to the fact that some of the active ingredient supply was missing or had been misplaced. Internal emails suggested that this was raised by the quality control manager on duty to higher authorities. However, this was not given due attention, either because of lack of knowledge, incompetence, or lack of interest in maintaining the highest standards on the part of the higher authorities.[10] The problem was not immediately reported to the regulators, to the suppliers, or to any other organization in the public sector. The staff on duty finally located the source of the problem. A barrel of active ingredient was indeed missing. This correctly identified the source of lower concentration antimalarial. This should have raised some immediate concerns. Such as where did the antimalarial barrel go? Once again, instead of a full-scale investigation, this problem was addressed by quickly replacing the missing barrel by procuring another one. No questions were asked either. The antimalarial supply reached its desired levels, and all problems were apparently resolved.

Not quite, it turned out. The antimalarial active ingredient barrel had, for some obscure reason, ended up in the wrong production line. It looked the same as the active ingredient for the antihypertensive, Isotab. Here, suspecting nothing out of the ordinary, the manufacturing system used this antimalarial instead of the antihypertensive, creating a deadly cocktail that claimed the lives of over two hundred people. The brazen lack of quality control, oversight, and negligence resulted in one of the worst public health crises in the sixth most populous country in the world.

The problem gained national and international attention. It was deeply embarrassing for the provincial government that was carving an image of clean governance and transparency. It was also personally embarrassing for the chief minister, Mian Shahbaz Sharif, who is known to be a workaholic and holds his cards close to his chest. There was no minister of health in the province. The chief minister himself was holding that portfolio. Questions were being asked whether he is spread too thin, as he was in charge of a dozen or more ministerial portfolios. Should someone, an expert in matters of health and medicine, be in charge of the provincial health affairs, as opposed to the chief executive?[22]

The issue lingered for nearly a year, in various courts, moving all the way up to the provincial high court and then eventually to the Supreme Court of the country.[23] The offices of Efroze Pharmaceuticals in Karachi were initially sealed but like other matters often seen in developing nations, other immediate crises took over. Issues associated with governance, politics, and, most importantly, national security overtook the news and people's attention. Those who were in charge of Efroze were never charged with criminal negligence. Questions about cronyism, political corruption, and backdoor politics were raised but nothing happened. Efroze operates today with full license all over Pakistan. Their website claims that they are in the business of producing quality pharmaceuticals.[24]

Whether that was a consequence of poor policy, cronyism, corruption, or the unavailability of appropriate field-ready technology that can test drugs regularly is unclear. What is, however, clear is that the drug manufactured, supplied, and provided to the unsuspecting consumers was poisonous and deadly and was not captured by the existing system in place. It therefore points to the breakdown of testing at several key points throughout the supply chain, including within the company. A WHO report, "Pathology of Negligence," points to numerous violations of safety codes and basic problems in the company that contributed to a culture of negligence that culminated in the disaster.[10] These problems included the presence of raw materials with finished goods, the contamination of drums with rainwater, among other violations of good manufacturing guidelines. Another dimension of the problem in Lahore is the lack of availability of reliable labs within the country to test drug quality, potency, ingredients, and efficacy. Sending tests to the United Kingdom is expensive and time-consuming and does not create a sustainable culture within the country. Would this problem be resolved if there was better, widely used technology in the country? Would it be resolved if there was better policy in place? Perhaps the most important question to ask is how unique was Efroze? If we are to turn back the clock and go back in time, would we be able to catch the errors and save lives? Or was it inevitable that something like this would happen, with perhaps more incidents in the future, until a major overhaul of the system takes place?

The incident in Lahore brings up several other concerning factors. First, an analysis of the incident should be made in the greater context of a society that is highly fatalistic and does not have a custom of investigating hospital deaths. A postmortem is done only in rare circumstances in a society that, for historical and religious reasons, wants to bury its dead as quickly as possible. One can also not deny the fact that many of the patients who died were already sick, and hence their deaths, while sad, may not have come as a complete shock to the relatives, who are poor and on the margins of society. There were some exceptions, like Ashiq Hussain, who seemed fine to the family, but he did not represent the majority of the patients. With the large number of deaths in hospitals in Pakistan, this story was unlikely to make the news except for one reason: the sheer number of people who had died in the same ward and in such a short period. The role of media that thrives on "breaking news" in Pakistan also contributed to the general public finding out about the story. While there are some good guesses, the exact number of people dying because of the scandal is still unknown (it ranges from 178 to 213 in various estimates), largely because there may be people whose condition worsened over the coming days, and they died later due to other complications. There may have been other people who died at home or in the poorer suburbs of Lahore who were never counted.[6] Thus, it is likely that the number is much higher than the one reported by the government, which itself has been inconsistent in reports.

The second problem seen in Lahore and elsewhere, is the complete lack of awareness or understanding about the nature of the underlying problem. The responses of the government, the press, and even some experts raise serious questions about understanding the nature of the problem. Pharmaceutical development, quality control, and drug testing are issues that are never discussed in the public domain. In general, scientific literacy is quite low in the country and quality control is not an issue that is often discussed. Conspiracy and science are mixed together and occasionally conspiracy theory and nonsense substitutes for science, sometimes even in science textbooks.[25] Homeopathic outlets are considered hospitals and little information about drug quality, activity, and the process of discovery is available to the public. With the influx of subpar products coming from abroad, particularly from China, people have learned to live with products that are not of the desired quality and that may fail routinely. The poor-quality issue is not just with drugs and medical products, it projects itself in all areas of consumer goods ranging from cell phones to winter heaters and all other kinds of household commodities. There is a general expression of "*do number maal*" (translated literally as a second-rate product, meaning something that is not of the desired quality), and the same applies to pharmaceuticals as well. So while there is no fundamental appreciation of the scientific process in drug development or how quality control would work, there is general acceptance of poor-quality products. So when the Lahore incident unfolded, there were many immediate reactions pointing to various conspiracy theories and fanciful explanations that had little to do with reality. There was also a proverbial shrugging of the shoulders, and some called it inevitable, because poor-quality products are not unheard of in the society.

Another dimension is to put the blame on the patient. This was particularly true for the subsequent case in Lahore about the cough syrup. In the Tyno case, the blame was squarely placed on the poor residents of the slum who were using Tyno as a recreational drug. This blame situation erupted for two reasons. First, there is indeed a group of people, largely urban poor, who use cough syrup and other sedatives as recreational drugs. Second, the government was reeling after the Lahore scandal and concerned about its image and wanted to save itself from a serious embarrassment. Consequently, it reacted quickly. No questions were asked about either the sudden spike in the numbers of the dead or drug testing. In a typical sensational manner that reflects the current news trends in Pakistan, the issue became political before it even had a technical component. The government made it clear that its hands were clean, even before any inquiry was launched. The government backtracked its original blame on the slum residents only after the family member of one of the deceased came forward and refuted the government claims about his relative being a drug addict. The media, which was not entirely supportive of the government, jumped at the opportunity and hence forced the government to start an investigation. The investigation pointed to a complex set of factors, including overdosing, but also did not rule out the possibility of contamination.

Fourth, poor-quality drugs have an impact on the patient that goes beyond consumption. The incident in Lahore was quickly followed by another incident in the northwestern city of Peshawar where tens of thousands of interferon injections, needed desperately by hepatitis patients, were deemed of inadequate quality. The chain of blame went up and down the procurement department and the local bureaucracy. A number of officials were either dismissed, placed on forced leave, or charged with criminal negligence. Large batches of interferon, numbering in the tens of thousands, were thrown away. Some contained nothing but dirty water. Because these injections were going to be given at local public hospitals, the sudden unavailability created a severe shortage and forced the already poor consumers to buy interferon at private pharmacies. This sudden and sharp increase in demand resulted in pharmacies increasing their prices quickly because the government has little control on pricing at private pharmacies.

The case of poor-quality interferon has lingered in Pakistan, reaching the highest court, where the owners of the pharmaceutical company, Pharmedic Laboratories, involved in the manufacturing and those involved in procurement got a stay against their arrest. In late 2015 it was reported that a steady level of supplies had reached the province to become available for the patients at the public hospital.[26] However, it is always uncertain whether the steady supply is sustainable or at the mercy of the next crisis.

The situation in Cape Cod, and in many other parts of higher-income countries, suggests a completely different and evolving challenge emerging from online pharmacies, Internet commerce, and peer-to-peer sales that go under the radar. In these countries, the lifesaving commodities are usually well regulated and not typically at risk, but the impact on health, from buying knock-off Viagra or a herbal supplement can still be substantial and the financial impact to the drug manufacturer is certainly noteworthy. Pfizer estimates that the annual cost of Viagra through black market and counterfeit trade amounts to millions of dollars in lost revenue. The exact impact on health is unclear and varies, depending on what the counterfeit actually contains. It can range from no effect to serious side effects that may require hospitalization. The trail of where these counterfeit drugs are made, how they reach the market, and who is behind them is varied and complex. Viagra may be the most commonly counterfeited drug sold on the Internet, but it is not the only one. Counterfeit Cialis, Valium, Xanax, and Lipitor are traded routinely online. Similar challenges exist about dietary supplements and herbal supplements that may contain nothing but paint or chalk. With the growth in Internet commerce around the world, it would be naïve to think that the problems are only in high-income countries. Recent studies and news reports suggest that online pharmaceutical sales are picking up in Russia, India, China, Brazil, and several other countries in Latin America, pointing to further challenges in maintaining drug quality.

From a big-picture perspective, the problems of substandard and counterfeit drugs can be classified into four broad categories. The first one is associated with regulation, policing, and consumer protection. Are countries and regulatory bodies complacent or complicit in bringing change and addressing the issues? Do the countries have strong-enough laws to prosecute those who are found guilty? This issue, while central to improving drug quality, is extremely complicated to enforce. The time between the first seizure of a shipment (or a batch of poor quality) and a trial can range from several months to over a year. Sometimes it takes nearly two years before the case comes before a court.[27] Often, after long litigation, the end result is a minor fine or a suspended sentence or a mistrial. In Kenya, for example, the fine after a first offense, proved after years of litigation, may just be a few hundred dollars.[28] For a company or a distributor that deals in tens of millions of dollars, this would not even register as a serious fine. The policing issue, which should start with manufacture and quality control at the development sites, is also complicated for many developing countries that do not produce a lot of their own drugs. Thus they depend on foreign companies that operate through a complicated framework of distributors and suppliers for their pharmaceutical supply chain. This results in the local government and police having little authority to impose strict penalties, besides license suspension, which is also a rare instance. The impact of the growing influence of China, with African nations in areas other than pharmaceuticals also affects the ability of the governments to impose strong penalties.

The inherent corruption in countries, unstable political systems, and regulatory bodies not having a clear, apolitical mandate complicates things.[29] For example, in Pakistan, the process of devolution of the ministry of health has meant that in the city of Karachi both federal and provincial inspectors operate side by side, with no interaction between them whatsoever. The federal and provincial inspectors may choose to procure samples from the same supplier, the same pharmacy, or a different one, and they are never required to coordinate with each other. Despite being a country of over two hundred million, Pakistan does not have a single standard operating procedure for procuring samples for testing. It relies exclusively on the desire and the whim of the inspector, who is often untrained in statistics or rigorous methods of sampling.

The problem of regulation is further complicated by the fact that the laws have not evolved with newer realities of commerce in the Internet and digital age. Thus the regulators are often unable to cope with the new dimensions of medicine procurement and use, and they may not have the tools or the mandate to deal with newer modes of trade, import, and consumption. With the growth of Internet pharmacies, complicated supply chains, globalization, and the prevalence of medicines that enter a country but are not regulated or even registered, ensuring patient and consumer safety is a very difficult challenge. This is further complicated by the fact that regulation is, in general, reactive and not proactive. The long debates in legislatures, in the light of other immediate and politically profitable issues, lead to regulation becoming a low priority. In many cases, there are no clear laws against substandard drug imports and sales, and the loopholes continue to embolden those with nefarious interests.

The second broad category of challenges is the physical security of the pharmacopeia. What if the drug supply was disrupted by a terrorist incident? What if the central medical stores are looted or become a victim of arson, as happened in Ghana? Here the issue is not simply that of regulation but also policing and providing the requisite security that is needed to safeguard the people against threats, internal and external. The impact of a contaminated or threatened supply chain is severe, and that can become a national security issue. The case in Accra suggests a fundamental compromise of the system and exposure to challenges previously unanticipated. Assault on the supply chains, or a severe delay in arrival of supplies, or a recognition that the supplies are not of the desired quality then lead to an acute shortage of drugs that are increasingly coming from abroad. In the case of a natural disaster, these challenges can significantly impact the health and well-being of a large section of the society. The interferon situation in Pakistan points to this challenge. Once it was determined that the currently available drugs were not suitable, and had to be taken off the market, the poor and vulnerable communities were disproportionately affected. They were no longer able to get the drug at the public hospital, the good private pharmacies had prices beyond their reach, and hence they were likely to procure knock-offs from unlicensed stores and further strengthen the hands of illegal traders.

The third category of challenges reflects the need for newer and better resources, both human and technological. There is a clear gap in adequate tools available, for the inspectors, regulators, and the justice system to carry out the necessary work to ensure pharmaceutical safety. The technologies that are currently available in the market often fail to deliver in a timely or comprehensive manner. Technologies in use in more technologically advanced nations are not easily transferrable to places in low- and middle-income countries. Gold standard technologies, such as high-performance liquid chromatography and mass spectrometry, are too expensive, costing hundreds of thousands of dollars, if not millions, in instrument costs (per instrument) alone. Having a large number of these instruments in the public sector is financially challenging for countries where the problem of poor-quality drugs is most acute. These costs are further multiplied by the need for expensive and often unavailable consumables that themselves rely on strong supply chains. There is, in general, a lack of trained professionals who are able to use these instruments effectively and are able to maintain them adequately. The challenges in physical infrastructure, such as grid power, clean water, air conditioning, and humidity and pressure control in labs, further add to the costs of the technology. Should countries decide to cobble up resources to build some of these labs that have the physical infrastructure, they are often unable to maintain them, leading to instruments going offline more often than staying functional. The lack of a local trained workforce to service instruments or fix them leads to instruments becoming useless and obsolete. The typical model of service contracts with manufacturers is expensive and hard to maintain. As these instruments age, it is harder for the countries to find companies abroad that are willing to service them, leading to a vicious cycle of disrepair that sucks financial resources on one side but does not deliver on the other.[30] The technologies currently used in many developing nations therefore are poor substitutes of what is needed on the ground.

The issue of technology goes beyond the absence of necessary equipment in central drug-testing labs. It also extends to adequate training and technological empowerment of inspectors and regulators. In the absence of tools that are portable and easy to use at the point of care, or the point of need, the best tool available to many inspectors is simply visual inspection. An unreliable tool, that is fundamentally flawed and subjective, thus leads to large gaps in drug-quality testing. Developing new technologies that are aimed at developing nations only is also neither simple nor straightforward. The incentive for the developers and innovators is not quite obvious, and often not financially rewarding, leading to gaps in technology that continue to grow with time.

The issue of resources encompasses people as well as instruments. Countries with large populations, in tens or even hundreds of millions, have often just a handful of inspectors, who are overwhelmed and undertrained. In Kenya, with a population of nearly forty-five million, the number of inspectors focusing on the issue of substandard and counterfeit drugs is fewer than twenty; in Pakistan, with over two hundred million people, the number of federal inspectors is around twenty-four.[31] In addition, there are few, if any, protocols or procedures to operate and coordinate within the country. The inspectors are neither trained nor qualified to do sampling, statistically and reliably. There are no regular training refresher courses that are comprehensive and up to date. With paltry salaries and little training, their sampling and methods are often incapable of fully understanding the depth of the problem. On one hand, there are glaring omissions in sampling, and on the other hand there is a disastrous level of redundancy in sampling. This leads to the wastage of resources in places where resources are most precious and rare. With hands tied behind their backs, the government agencies and regulators have little ability to police or enforce best practices across large swaths of the land.

The final category of challenges is the lack of awareness and understanding of the complexity of the problem. A clear illustration about limited awareness is the analysis of general discourse on the issue. The general discourse follows two main themes. The first theme is exclusively focused on counterfeit drugs, where there is a breach of patents and the focus is on intellectual property, and hence the pharmaceutical industry is acutely concerned about its brand protection. The other angle of the conversation is focused on the intentional falsification of drugs, initiated by those who are interested in making a profit from naïve and unsuspecting consumers. While both of these areas are very important and highly relevant for public health, they do not paint the complete picture, particularly in developing nations.

The awareness of public health challenges posed by substandard or poor-quality medicines is often appreciated only by public health professionals, and rarely by the general public, the media, or the politicians. There is also a lack of conversation and understanding between public health professionals and drug inspectors, regulators and policy makers.

This lack of awareness results in misinformation, hyperbole, and misrepresentation of the problem, particularly in communities that are prone to conspiracy theories and whose residents suffer from poor, if any, scientific literacy. The origin of the problem might be closer to home, but the lack of knowledge tends to blame it on sinister elements among the enemies of the state and on a global mafia operating in the shadows. This can lead to impulsive reactions and panic in ways that do significant harm.

The lack of awareness among politicians and the media often simplifies the problem, or it paints a picture that is inaccurate, dangerously incomplete, and fragmented. The issue of awareness is also present among the research community of engineers, technologists, and innovators. This leads to an asymmetric focus on problems and lack of development of new tools and technologies. The issue affects clinicians, many of whom do not fully understand the issue of drug resistance. Pharmacy students are not entirely immune either, largely due to lack of exposure.[32] The curricula in pharmacy in most countries, where the problem of substandard and counterfeit drugs is acute, fail to mention the challenges of these drugs, do not clearly identify the bottlenecks or gaps in knowledge or technology, and rarely encourage students to develop any interest in creating new and problem-oriented solutions.

And it continues up the chain. It leads to little interest by funding agencies, including those in developed nations, to address a problem that makes upward of $70 billion annually. A lack of comprehensive understanding and no appreciation of the multifaceted nature of the problem mean that private institutions and investors have no interest to invest in new technologies or initiatives. Home-grown solutions that are robust and scalable, culturally cognizant and scientifically rich, are less likely to get the backing of investors and institutions due to their poor understanding. This increases the dependence on importing solutions and approaches, many of which are inherently unsuitable for local context.

The cities of Luanda, Cape Cod, Lahore, and Accra are not the only ones facing a crisis. Globalization and international commerce, politics, and development mean that a crisis in a faraway land can be felt closer to home in the suburb. Resistance to frontline therapeutics affects communities in the highest-income countries as well as the weakest economies.

The crisis of bad drugs is a challenge that is both inherently global and uniquely local.

2

The Life of a Pill

In the fall of 2012, after attending the annual meeting of the American Society of Tropical Medicine and Hygiene in Atlanta, I visited the lab of Professor Facundo Fernandez, a chemistry professor at the Georgia Institute of Technology. Facundo is a soft-spoken Argentinian who is an expert in a technology called mass spectrometry. Mass spectrometry is a highly precise mechanism to identify the chemical constituents of a product or an unknown sample. An impressive device, both in terms of its function and in terms of its cost,* a mass spectrometer can precisely measure the masses of various chemical components within a sample. It is quantitative and accurate even with trace amounts of materials. It is a gold standard instrument when it comes to various fields of analytical chemistry.

Facundo is a leader in the field and, for a while, has been working with colleagues at the Centers for Disease Control and Prevention in Atlanta. He collaborates with researchers all over the world to test drug quality and uses his tools to quantitatively determine exactly what is in the samples collected from across the globe, mostly from the frontlines of global public health.[1,2,3] As he gave me a tour, I was like a kid in a candy store, in awe of all the gadgets in his lab.

* A mass spectrometer can cost upward of a million dollars, and that does not include the need for trained and experienced personnel.

In the last room that we entered, he showed me neatly arranged boxes of samples of drugs, collected from Africa and Asia, all waiting to be tested in his lab.

Facundo showed me results of various tests that he had been doing lately. Of the ones that he showed, there was one that was particularly intriguing and caught my eye. It was the test of a malaria tablet collected from Southeast Asia. The tablet was coated with an antimalarial on the outside, but the inside was acetaminophen, the key ingredient of Tylenol and Panadol. Clearly a fake drug, it was meant to deceive not just the patient but also the inspectors who may just scrape off the top and analyze it, and never find that the core of the drug was not what it was supposed to be. Unsuspecting patients or caregivers, who would consume it to cure malaria, would see a decrease in the fever and would feel that the drug was working. But the relief would only be temporary since the drug had only a minimal amount of antimalarial. The consequence of consuming such a drug would be disastrous.

This begs the question, of not just the crime that is committed, but the loopholes in the system that allow for this crime to be committed. The question is simple: Where in the life of the pill can the system be breached with such ease?

Pills, syrups, and capsules have unique shapes, sizes, and even tastes. There is a mental image, in all our minds, of what Tylenol looks like, or why Advil should not have a strong taste, or how to best help children swallow aspirin. The final shape, form, and taste are part of a long, multistep process. Any person exploring the process and implications of counterfeiting, or falsifying a drug, should wonder where in this technical process does the system fail. Is the system prone to failure? Is it by design porous and weak, or are the counterfeiters and falsifiers so smart that any system, no matter how tightly regulated, can in fact be broken?

So where does a pill actually come from? How does it begin its life and its long journey? Who makes the pill? Why do all tablets of a particular brand look the same? How does a tablet take a peculiar round pancake shape or that oval egg shape that we always associate with it? Mundane as they may sound, these questions are actually the foundation of both understanding the process and the nature of safety checks meant to protect the public against bad drugs.

The question of pill manufacture and associated counterfeiting, falsification, or low-quality production can be divided into three main parts. First is an analysis of how the pill is made from a technical standpoint, and why, and how, it may fail to adhere to the quality assurance standards. Second, from a big-picture perspective of global markets and big businesses, is how this process operates, and how it fails in the case of substandard products making their way to the market. The third is about how these two processes, the technical and the commercial, engage and interact and subsequently give rise to both successes and failures in the increasingly decentralized process of drug production.

The process of making a pill or a syrup, in modern times, is a far cry from what it used to be.[4] Gone are the days when the chemist would grind up the ingredients, bake things at a known temperature for a prescribed period, mix things again and again to get the right consistency and produce a final product. In doing so, the chemist, or the compounder as the person was called in certain parts of the world, was also in charge of the whole process. The chemist was the person whose biggest asset was his or her knowledge, experience and the trust he commanded.[5] The chemist often had some interns and apprentices, but ultimately he (and in those days, it was mostly men who were chemists) was in charge of the operation, the secret recipes, and the integrity of the process. He had most likely trained under another senior chemist before he started his own business. He also had a name and a face; he was a member of the community and was someone whom the local people could personally engage with.[6]

Times have changed. There is no longer a single person or a queue outside the chemist's shop. The operation of baking and milling, mixing and grinding, is no longer visible to the patient, or dependent on a single person. The process has moved away from the patient to labs and factories with restricted access. There is no longer a person whom you could chat up while he mixes potions in his apothecary. There is no longer a carefully guarded recipe book, though there still are plenty of trade secrets that are highly profitable. These days, the process is varied, complex, and multifaceted. It involves machines and standard operating procedures (SOPs) and multiple units operating in an assembly-line mechanism. While the process toward the end may often be similar across companies, where it all starts is far from common. It has also evolved, and continues to change regularly.

Drugs can be new discoveries or old, accepted formulas. The process of finding a new drug often starts by a discovery, in an academic lab with a graduate student or a postdoctoral scholar, far from any pharmaceutical industry. If we were to take a snapshot of the life of our proverbial postdoctoral scholar or our graduate student, we would find that he or she often is not looking for a blockbuster drug. Instead his or her discovery may be about something completely different. It may be about how something functions or malfunctions at the genetic, molecular, or cellular level. He or she may not even be engaged in pharmaceutical research altogether. Instead, the question at hand is often about fundamental function and process, not about finding the elusive magic potion. Most likely the funding for research for this work would also be coming from public sources as private funding for research still makes a small fraction of the total investment in academic research.[7]

Sometimes she or he may find a molecule that shows unusual properties or promise in shutting down a function inside the cell, or enhance the process by which the cell may commit suicide, or send a signal that influences other cells in the neighborhood. Or the student may stumble upon a process that was previously unknown or underappreciated.

With computing becoming more ubiquitous, even the process of discovery in a cold, lonely lab has undergone a major transformation within the last decade.[8] While many discoveries still happen in labs with white lights in gray buildings, the discovery does not need to be made in test tubes or in a dark room under a microscope. More and more, we get our Eureka moment through a powerful computer program. The computer program may be focused on understanding a fundamental biological process, and in doing so may uncover new insights, or discover a new binding sight on a molecule that is a potential target for a potent therapy. The computer programs may also utilize an informatics or a big data approach.[9] These informatics methods are able to look at hundreds of thousands, or perhaps even millions of molecules and compounds, match them against a possible target to see whether a molecule may act as a key to lock or unlock a process. The computer program then may search for better and better keys against this lock, screening them against tighter guidelines and eventually may find a small subset of potential drug molecules. It can also be run simultaneously on many computers, and the parallelizing of these tasks makes the process more efficient. Testing this small subset is significantly cheaper and more efficient than screening every possible molecule in the lab.

This hybrid computational and experimental approach, where the computer guides the experiment, has taken off in the last decade. This process of "rational drug design," as opposed to a brute force strategy of throwing everything at the target and seeing what sticks, is being used increasingly while our knowledge of chemistry, physics, drug action, and ability to screen existing molecules is cutting down on time, cost, and even animal experiments.

The road from discovery to potential drug candidates and eventually to a final drug molecule is long, winding, and full of risks of failure. Our knowledge of biology and disease is far from perfect, so the process of taking a potential drug to the next stage is not an exact process and hence highly prone reaching a dead-end. Yet this early stage process is necessary and the bedrock of our pharmaceutical arsenal.

The successes and failures of the lab research are far removed from what we may see at our local pharmacy.[10] The success rate from early stage discovery to a final pill is very small, almost minuscule, yet it is the first step in that long and winding road. At each stage there are tremendous costs, of human hours, of resources, and of wastage through failures. It is often this long pathway that the pharmaceutical companies point to as an argument for drug pricing. There is some merit to this argument, though there is also a sizable portion of the investment from the pharmaceutical companies that is spent on advertising and marketing that sometimes is significant, when compared to the overall budget spent on actual drug development.[11]

There was a time when nearly all of the discovery phase was happening in-house at pharmaceutical companies. In the last couple of decades, that has started to change with an increase in research activity outside the pharmaceutical companies.[12] This is driven in part by a desire by the pharmaceutical companies to cut costs through outsourcing research and engaging multiple groups at various stages of drug discovery. These days, a lot of early stage discovery is happening at the universities and research institutions, funded largely through grants from the public sector and occasionally through support from private foundations and pharmaceutical companies. This "farming" of the early stage process outside the company is a means to decrease the cost of discovery by outsourcing fundamental research away from their own companies. Because the cost of research at universities is often lower than in-house research, the companies are able to stretch their funds and encourage the pursuit of multiple options. Additionally, because the discovery process has a high chance of leading to dead ends, through outsourcing the cost of failure is also decreased. Combined with this outside farming approach is the strategy by pharmaceutical companies to invest only in research that has already shown some promise. They do that by licensing promising patents or engaging at a slightly more mature stage of discovery once some of the most basic hurdles have been overcome.

While discovery is a central tenet of research-based pharmaceutical companies, a drug store in a developed country would carry drugs from both the research-based companies and companies that produce generic drugs. Over the years, there is an increasing number of generic medicines that are available in the pharmacies in the developed countries. In developing nations with limited resources, this number of generic drugs is substantial and continues to grow.[13]

The process of making drugs at generic companies is both similar and different than what happens at a large research-based pharmaceutical company.[†] It is different because there is no discovery phase and hence the R&D component is significantly smaller (or nonexistent) compared to the research-based companies.[14] The generic companies focus on off-patent products that are no longer protected by their original developers and manufacturers. This does not mean that the lines between

[†] In many publications, research-based pharmaceutical companies are also called multinational companies, or MNCs. But since a number of generic companies are also operating internationally, we use the term research-based, as opposed to multinational.

research-based pharmaceuticals and those that focus on generics are clear. In fact, there are constant battles between the two groups about "ever greening" of patents, a process by which a small change in the original product would make it into a "new" product and hence eligible for a new patent for another couple of decades. The generic companies often argue that this practice is intentionally misleading and carried out only to stifle competition. They say that the practice results in no new discovery or any major change in the drug's function. The large research-based pharmaceutical companies dispute this claim and argue that the change is real and leads to measurable differences in product performance.

There are, however, similarities between large, research-based, and generic companies as well. The process of making a final pill from individual components and mixing and milling them together is fairly similar.

The discovery phase, proved and validated after a rigorous set of tests, moves to trials, which itself is a long, expensive, multistage process involving not only the pharmaceutical company, but also hospitals, doctors, ethicists, and regulators. The set of trials may include testing on human cells cultured in the lab, samples of human tissues donated from patients, and then on live animals. Drugs that show promise are then moved forward for human clinical trials, a process that is multistep, carefully controlled with extensive SOPs, managed by multidisciplinary teams and often funded privately.[15] Over the years new companies have emerged that specialize in conducting clinical trials and ensuring that integrity of the process is maintained and not influenced by one side or the other. The goals of clinical trials are patient safety, drug efficacy, and minimization of dangerous consequences or taxing side effects.[16] Because of the emphasis on science and discovery, the carefully choreographed and cross-validated nature of the process of clinical trials, the chance of counterfeiting or a substandard drug moving forward through clinical trials is virtually nonexistent.

The success of the clinical trials ushers in a new and promising stage in the life of a pill. It goes from an idea, a discovery, and a promise to a tangible product with immense potential stored inside.

This is also where the real chance of counterfeiting, or that of manufacture of a poor-quality product, starts.

Once the drug is approved by the regulators, whether it is brand-new or a drug for which the approval was granted decades ago and it is no longer covered by the patent, the process of manufacturing for commercial use can begin. The key goal of the drug manufacturing process, from the perspective of the industry, is that it needs to be efficient, cost-effective, robust, reliable, and error-free. Tens of millions or perhaps billions of tablets are made by one company; therefore, the process of manufacture needs to be highly reproducible with minimal variation between any two pills in size, shape, texture, or any other physical features. The level of precision in reproducibility requires that tight protocols are followed in excruciating detail and that quality instrumentation is available at all times.

There was a time when the end-to-end process used to be completely in-house and tightly controlled by a single manufacturer. This is no longer the case. These days, the various components of the manufacturing process are being carried out by companies that do not interact with each other directly.[17] Often, they are not even on the same continent. The processes of sourcing of materials, putting them together in various formulations, and even packaging are being done in disparate locations, and not just for generic manufacturers. It is also true for large research-based pharmaceutical companies. US government data suggest that more than half of pharmaceutical active ingredients going into the final product of the pill that are sold in the United States are coming from India and China.[18] Furthermore, the company that may be providing the active pharmaceutical ingredient to a US company itself may be sourcing various products from not only multiple sites within the country, but also outside its borders. The entire process is extremely distributed, and has at times been compromised to produce disastrous results.

The process of tablet making is a multistep process that requires each batch to have similar physical (e.g., color, smell, hardness etc.) and chemical (performance inside the body, solubility, etc.) properties. Each tablet needs to have certain common characteristics. For example, the particles that make up the powder of the tablet must stick together or "lock." Otherwise, the tablet would disintegrate. The tablet should not be overly sticky, which would lead to sticking to the packaging or having an undesirable texture. On the other hand, a highly dry product is also problematic. An overly dry product would also lead to disintegration. Similarly, each tablet, among millions and millions produced, needs to be of the same weight and same shape, not only within the same package and the same box, but when compared against any other sample of that particular brand. The average weight of a tablet may be just half a gram or less, which means that a tiny variation of even a tenth of a gram would be unacceptable.

Within this long interconnected chain of drug manufacture, each step is called a unit operation.[19] The idea is that by separating each step individually, it is easier to ensure quality control for that step. This also allows for development of unique protocols for each step in order to ensure quality maintenance. Additionally, by separating the batch operations, it is also possible to outsource them and have different groups within, or outside, the pharmaceutical company take charge and be held responsible for that particular unit operation. Groups and companies that have developed expertise in a given unit operation can then market themselves to various companies looking for that particular set of skills. While this allows for these unit operation experts to create a viable business model, it also means that errors on their part would be multiplied across various companies that are sourcing products from them.

The final drug available in a packet, box, or bottle contains more than just the active ingredient. It has additional components called excipients.[20] These excipients are necessary for a variety of physical and chemical functions of the drug. These features include better solubility of the drug inside the body, its ability to absorb inside the gut, and ensuring that the tablet integrity is maintained for the period before its expiry. The excipients are also important for the final appearance, taste, and smell of the drug, and they are added to increase the appeal to certain demographics, such as children, or to enhance the relative time for action such as gel coating that reduces the time for absorption.

The excipients and the active pharmaceutical ingredients often do not mix automatically due to their different chemical properties. As a result they need to be "glued" together through a pharmaceutical-grade adhesive called a "binder."[21] These mixtures of active ingredients and excipients, glued together, are then converted to make granules. The granules then go through various stages of milling, processing, and drying to create drugs that have the right consistency, texture, and homogeneity across the entire tablet. Circumventing this stage leads to an inhomogeneous tablet, which is often the hallmark of a counterfeit or a substandard product, such as the one Facundo tested through his mass spectrometer. The product coming out of multiple stages of milling and drying is then compressed using a tablet press. The tablet coming out of this stage also needs to have a consistent weight. This nearly final product is then coated with a thin film that can give it a specific color, texture, and taste, as well as make it possible for it to be handled easily within the package and also by the consumer.

Not all drugs that we see on the market have a solid tabletlike form. Many are available in color-coded capsules. These capsules are basically drug-carrying vehicles. Making a capsule (a process called encapsulation) is inherently similar to that of making a tablet, except that in the final part the process, the components of the drug, and excipients are added to a two-part capsule or a gel coating. The goal of this process is to increase absorption and decrease the total time it takes for the drug to act. Capsules can also contain liquids and therefore have an added flexibility in terms of the components that it can carry.

The drug after milling, drying, and pressing is nearing the end of its manufacturing life and is almost ready for packaging and shipping. But before it can be sent for distribution, the product is printed with the name and brand of the manufacturer using a pharmaceutical-grade ink. Each tablet is then put in various kinds of packets, sachets, boxes, and bottles that make it to the local pharmacies. The packets are designed to save it from disintegration and protect it from the weather elements.

Shipping the drug is based on a number of factors, including its chemistry and sensitivity to temperature, moisture, sunlight, and other outside elements. Regardless of the chemical nature of the drug, extreme temperatures fluctuations can seriously affect drug performance.

Given the the various steps involved, one can imagine that a lot can go wrong, and it sometimes does. Even companies that have been in operation for decades or more, struggle with consistency and quality maintenance.[22] As a result, strong quality assurance and quality control measurements are in place to maintain the integrity of the process. This requires testing the drug at various stages of development, including analyzing the products at various points and testing random samples for both their chemical compositions and their physical properties.

Some of the most common and obvious signs of a product not being of the desired standard are physical characteristics that are obvious to the naked eye. These characteristics range from something that may look sticky, swollen, discolored, increasingly fragile, chipped, overly soft, or damp. These are obvious telltale signs that can raise suspicion, but not all aspects of poor quality are obvious and hence require more rigorous testing, which needs reliable equipment and technical expertise. Absence of either, or both, is often where the problem of substandard drugs starts and snowballs as it moves through the system. The Pakistan drug scandal of 2012 is a case in point, where the problems with the drug could not be spotted with the naked eye. However, early testing and more competent technical staff at the company could have acted early and saved the lives of hundreds of people.

The process of pharmaceutical manufacture can be divided into two broad categories. The first is the actual process of drug making, as discussed above, which includes milling, printing, pressing, and gluing. The second aspect is the big picture of how the various parts going into the manufacturing are obtained, sorted, and stored and how these disparate components are put together before they are moved further along the chain, all the way to the consumer, which may be an individual at home, a clinic, or a hospital.

The process of drug manufacturing requires sourcing of materials needed at each step, manufacture of semifinished products in various locations, and the final stage of integration. For example, one can procure the active ingredient from a supplier in rural China, process it and add excipients in Beijing, add the capsule coating in Spain, brand it in one city in Panama, and sell it in a number of locations in and outside the country. The success of each step requires two fundamental components: a good input from the previous step and quality control in that particular step. The entire process is interlinked with compliance with local and international law, trade rules, regulations, and process validation at each stage before the product moves to the next step. Documents verifying compliance and scientific tests are needed to ensure that the process in the previous step was done in an appropriately rigorous manner and to leave a paper trail in case something goes wrong.

While, in principle, each stage would be governed by a similar set of laws and policies, and would act in a manner that allows for tracking of errors, in reality it is far more complex and error-prone. Some errors are genuine mistakes, and some are a consequence of human negligence while others are due to criminal activity. For example, if an active ingredient or an excipient is not of the desired quality, and the documents describing the scientific tests analyzing this component are fabricated, the problem is going to be passed downstream. Increasingly, various components, in raw or semifinished form, are now being sourced by third parties that compete internationally and promise to provide the best price. These third parties may not have the necessary expertise themselves and may be sourcing their product from somewhere else. As the chain of responsibility continues to hop from one source to another, the chances of failure increase. In case any one of the sourcing parties provides a tainted material, and uses fabricated documents to authenticate, the final manufacturer, relying on the good faith of its sources, may not be able to catch the compromised raw materials.

This hypothetical scenario, unfortunately, occurs frequently enough in the real world to cause serious damage.

In 2006, nearly four hundred people, many of them children, died due to drug poisoning because of a contaminated cold medicine in Panama.[23] The problem did not originate in Panama, where the final product was made. Instead it started thousands of miles east of Panama at a company called Taixing Glycerine in Hengxiang, China. Taixing Glycerine had made no direct contact with the manufacturer in Panama. Rather, the company in Panama had bought the active ingredient through a Chinese state-owned distributor. The Chinese state-owned distributor itself had gotten the active ingredient from Taixing Glycerine. Taixing Glycerine had sourced its glycerine, a key ingredient needed for the final active ingredient, from a forty-something-year-old individual named Wang Guiping, who was interested in trading chemicals.

Mr. Wang had previously been a tailor but got interested in selling chemicals for pharmaceutical manufacture. He forged the license of his product as well as chemical tests on the product. He needed to provide glycerine, but glycerine was expensive, so he looked for a cheaper substitute. Instead of using glycerine, which was the required product, he substituted it with diethylene glycol, which was an industrial chemical used in antifreeze, and not a pharmaceutical-grade ingredient.

Mr. Wang made the switch based on what he saw in a chemicals book, and he never tested or tasted (as he had done with other chemicals) what he sold to a number of manufacturers. He supplied his product, which was poisonous, to a number of Chinese pharmaceutical companies that were interested in buying glycerine for their pharmaceutical products. Among these Chinese pharmaceutical companies that bought fake glycerine was the Taixing Glycerine factory, the one that provided the glycerine syrup to Panama. No one at the Taixing Glycerine factory tested the material provided by Mr. Wang. Instead they believed and relied exclusively on the fake certificates that he provided with his supply.

On their way to Panama, dozens of barrels of this toxic syrup that contained an industrial grade antifreeze passed through several ports and checkpoints. From Hengxiang, it went to the

FIGURE 2.1.

Recent years have seen a number of global crises originating in China. Poor manufacturing, middlemen, and a lack of regulation has impacted consumers both within and outside China. The figure is taken of a counterfeit drug-manufacturing site that produced high-profile drugs like Viagra (and other drugs) for unsuspecting consumers in Europe and the United States. Reproduced with permission from Pfizer.

state-owned trading company, called CNSC Fortune Way. CNSC Fortune Way had started as a trading company dealing with supplying goods and services to Chinese personnel and businessmen overseas. In the early 2000s this trading company expanded and started supplying pharmaceutical ingredients as well. From CNSC Fortune Way the barrels went to Barcelona, Spain, carrying a certificate saying that the barrels had 99.5 percent pure glycerine. From Barcelona, the shipment moved to Panama. All along the way, the licenses simply copied the certificates provided by CNSC Fortune Way. These licenses did not even include the name of the original sourcing company or the person who provided the supposed glycerine. Those were discovered much later after the authorities started investigating the disaster in Panama.

As these barrels reached Panama, there was no reason to suspect any wrongdoing. All the certificates and the names of the handlers seemed to be in order and pointed back to a legitimate, Chinese state-owned company. That was all that was required at various checkpoints.

At no point, from China through Spain to Panama, was any testing done. It is also not clear whether any tests were ever carried out in China. Had that happened, the quality assurance staff would have figured out that there was no glycerine in the product.

The impact of poor sourcing and lack of tests was not only deadly for unsuspecting patients in Panama, but also for dozens of people in Guangdong Province and in Sichuan Province, all supposedly dying from the tainted active ingredient provided by Mr. Wang alone. No one knows the exact number of casualties and any guess would be an underestimate. This is because people who ended up dying were already sick and hence their sudden passing was not an immediate cause for concern. Additionally, their loved ones may bury them before any investigation formally takes place.

The sourcing problem, and the lack of legal oversight in testing raw materials, is a messy business and often a situation where different government agencies point to a lack of mandate or enforcement authority. Upon pressure from abroad, the State Food and Drug Administration of China tested the glycerine at the Taixing Glycerine factory and found that there was no glycerine in any of the products being sold or in any barrels that were labeled as having glycerine within them. All barrels only contained the same industrial grade antifreeze, diethylene glycol.

The case, however, was not simple. The Taixing Glycerine factory actually had no license or authority to make any pharmaceuticals whatsoever. This meant that their activities fell outside the legal mandate and purview of the State Food and Drug Administration of China. The Chinese FDA was therefore unable to move the case forward, as their role was to be the regulators and guardians of drug manufacturing. These breaches of the law had to be dealt with through new criminal probes, with their own bureaucracy, and not through the State FDA, which was told the crime was off-limits given their mandate.

The problem seen in Panama, China, Pakistan, and elsewhere, where the raw material itself is counterfeit, substandard, or falsified, requires cooperation and resources from all parts of the supply chain. However, this cooperation from various entities in the supply chain requires technical capacity, regular surveillance, and procedures that depend on active testing during sourcing of chemicals and raw materials. The cost of this oversight is high and can slow down the process significantly.

The problem of poor sourcing cannot only be blamed on the challenge of global supply chains. While that challenge is real, it is one of the many challenges associated with poor manufacture. Poor-quality active ingredients do not just enter the system due to systematic corruption alone. They also enter because the manufacturers, in order to keep the costs low, procure their active ingredients from the cheapest source, which is often unreliable. That, combined with issues such as poor water quality during manufacture, leakage and unsanitary buildings, absence of technically competent staff, equipment that is not fully calibrated, and other issues such as power loss during manufacturing, can lead to a product that fails to meet the minimum quality standard.[24] The outcome may not just be an ineffective product: It might lead to a product that is outright lethal.

Just as the final drug has a finite life, providing users with an expiry date before which the drug should be consumed, so do the ingredients that constitute the drug. Some ingredients are sensitive to light and humidity; others start to degrade in heat. Their safety and stability are just as paramount as that of the final product. Similarly, just as the drug may have certain requirements in terms of storage, so do the ingredients. A final product, even with the right ingredients, may lose potency due to exposure of one or more of its ingredients to extreme temperatures or humidity. The whole process may start on the wrong foot because of the quality of one of its ingredients.

Manufacturers, both generic and those making research-based pharmaceuticals, often make multiple products for multiple diseases and have several operations and processes taking place at the same time. The chance of cross-contamination, in an environment that is not well regulated, or is prone to malpractice, across various processes is therefore very real. This problem of cross-contamination was the issue that led to the Punjab Institute of Cardiology crisis in Lahore in 2012.

During manufacture, it is not only critical to keep the ingredients pure, but it is also equally important to have oversight of the process. Regular checkups on how and where to store ingredients is a key element of quality assurance. SOPs also require that raw materials and finished products are not to be mixed, things should be labeled appropriately, and chances of cross-contamination are to be minimized.[25] Yet this is not always straightforward. In developing countries, many workers are daily wage employees, working on a contract with a third party, and may not even be literate. As a result, their lack of training and knowledge can lead to an increased chance of cross-contamination. Many powders look the same and may not have any particularly distinguishing physical feature or particular smell. A powder that is used for blood thinning for patients of hypertension or cardiovascular ailments looks not too different from an antimalarial that is used for a completely different group of patients. While accidents are bound to happen, lack of oversight or ignorance can lead to deaths. In Pakistan's case, the barrels got mixed up and the absence of one barrel from the assembly line was not taken seriously enough.

To mitigate these factors and to increase the overall quality, there is pressure on manufacturers in developing countries, that governments and regulators from higher-income countries can apply, but not much. A recent trip by the head of the US FDA, Dr. Margaret Hamburg, to India showed just how difficult the task of changing the culture of quality can be.[26] The visit to Indian pharmaceutical manufacturers showed glaring problems at a number of Indian plants. Newspapers both in the United States and India picked up on the story of the visit of the FDA's high-level team to India, with improving the quality of drugs as a top agenda item. With India providing nearly 40 percent of all generic drugs in the United States, the impact of poor-quality manufacturing and lack of quality control is of significant concern to the US FDA.

The US FDA's increased emphasis on enforcement and quality has met with both scorn and anxiety. There is worry that the US standards are too stiff and too difficult to achieve for India, and compliance would come at a serious cost. G. N. Singh, India's top drug regulator, remarked that if the Indian regulators and Indian manufacturers were to follow the US laws and guidelines, there would not be any more drug manufacturing in India.[27] While India continues to provide a large share of the generic pharmaceuticals in the world, particularly in developing nations, and there have been a number of incidents about poor-quality medicines originating in India, not all Indian manufacturers are known for substandard products. Cipla, one of the largest manufacturers in the world, with billions of pills made annually, is well known for its strict quality standards and FDA-approved manufacturing plants, and it exports around the world.[26] Still, there is increasing pressure on India to improve its quality standards and regulations, but it is unclear what can be done in a short period of time.[28]

It is easy, and perhaps convenient, to argue for increased oversight, better technology, and more tests. In theory all of these things should happen, but in reality all of these interventions cost time and human resources. Testing at every single point in the supply chain is simply not viable or feasible for any country. The cost of such an endeavor would be significant. Additionally, it remains unclear if all governments and companies are equally interested in absorbing the costs of these additional processes. Requiring extensive additional testing may force the pharmaceutical company to pass the added cost down to the consumer, which can lead to public outrage, especially for essential and lifesaving medicines. Similarly, added oversight from the government would also require a major investment in human resources, which may not be a high priority for cash-strapped governments. Even if a government was to somehow manage additional resources, in low-income countries regular testing may not even be an option due to lack of electricity, trained capacity to maintain instruments, lack of grid power, and the high cost of consumables that can run in the tens of thousands of dollars per year for every single one of the high-end instruments.

From a political angle, forcing compliance on a major manufacturing country is also not straightforward. While the United States, as a major recipient of Indian drugs, can exert some pressure on the Indian pharmaceutical companies to improve quality, many of the other consumers of the Indian pharmaceuticals may not have the same political or financial leverage. India is a major supplier to many African countries that depend on Indian products to maintain their public health system and continue their fight against deadly infectious diseases.[29] First, these countries have much smaller markets, and are economically or politically weaker than the United States and do not have the same international clout to exert pressure. The business relationships of these countries are also not necessarily bidirectional, and they rely much more on Indian imports than exports to India. Second, if Indian manufacturers were to adopt much tougher criteria for manufacture and quality control, the added cost may make the drugs beyond reach for these countries. While this would be difficult for developing countries to sustain, this will also have an effect on Indian manufacturers and suppliers. The increase in price of the Indian drug would open up the possibility of other manufacturers from other countries to bring their products, which may be cheaper, and may not be of the desired quality. From an Indian company standpoint, it is difficult to create and maintain two independent product streams: one for the United States or similar markets and one for the developing world market in Africa and Asia.

The approach therefore cannot come from a single one-shot solution, or a top-down approach, to force the Indian manufacturers to improve their quality. Instead, a multiprong solution, one that provides tangible improvements at all points in the chain, is needed. This would require increasing awareness and an increase in the number and the training of the staff in both the private and the public sectors. The system strengthening should happen at both the manufacturer's end and that of the receiving country. Small changes at all points are much more likely to make a sustainable impact rather than a drastic change at the final point.

The system strengthening should start with the regulators and their capacity, both human and technical. Most countries do a poor job in having enough regulators or inspectors. While India produces over $15 billion annually in exporting drugs, the Indian Central Drugs Standards Control Organization, the national regulatory body, has a staff of just over 323. In a country of over a billion people and the pharmaceutical industry being one of the major contributors to national commerce and export, the size is under 2 percent of the size of the US FDA.[30]

The problem in other countries, from Pakistan to Kenya, is even worse. In Kenya, a country of nearly forty-five million people, there were fewer than a dozen drug inspectors in 2014.[31] In Pakistan, which produces a significant number of medicines for its over two hundred million citizens, the number of inspectors is a few dozen. The problem is not just in the number of staff members, or their severe lack of technical competence, but also in the archaic law that they are supposed to enforce. The drug act of Pakistan is over forty years old and has a large number of loopholes that are often exploited by the pharmaceutical sector.[32,33] The newly created Drug Regulatory Authority of Pakistan (DRAP) is still unclear on what its mandate actually is, and since its creation it still fights all kinds of political battles about its very existence and has remained largely ineffective in enforcing quality standards.[34]

In most places in low- and middle-income countries, where the national regulatory procedures demand regular oversight, testing is largely focused only on the final and finished pharmaceutical product, which itself happens in an ad hoc manner.[24] Even if it were to happen perfectly, it would not necessarily improve the quality of the system overall. Testing the finished pharmaceutical product would simply catch the bad drugs. By not focusing on testing the quality of the ingredients or routinely verifying the process, a drug company could continue with making bad drugs until it is caught. With the system being porous and the laws not always being enforced, the act of getting caught is rare and is unlikely to change the overall outcome. What is needed is therefore not only an increase in the number of people who serve as inspectors and technicians in various drug-testing labs but also improvements in testing protocols. These protocols should be based on efficient sampling methods that are able to quickly screen the drugs and send the suspicious ones for further testing and analysis. The protocols for testing also need to reach to manufacturing and early distribution, and not just the final product. Current laws in many countries, including major economies and contributors to the global market such as China, require the testing of either the final product, or the active ingredient, but not of the process itself or of the quality of other excipients.[35]

Another loophole widely cited for poor manufacture is associated with import licenses of chemicals and raw materials. Materials for pharmaceutical manufacture require a higher quality and a higher scrutiny. However, similar or identical materials can be imported under a chemical license for industrial manufacture, which do not need to be of the same quality and do not require the same scrutiny. Once the material is inside the country, it is essentially free for all to be used as they see fit. Because the industrial materials may be chemically similar or identical to the pharmaceutical ingredient (just not of the same quality), they are often substituted for pharmaceutical manufacture, resulting in dangerous products that are not fit for human consumption.[36]

Drugs are also not just created for human consumption alone. A high volume of drugs are also needed to ward off infection and keep animals healthy. This ranges from vaccination to antibiotics, many of which are used extensively in countries that have a major dairy or meat industry. While the drug manufacture may be broadly similar in those cases, the regulatory framework, testing frequency, human resources to regularly test, and the incentive to test in developing countries are often lacking. The impact on rural livestock, particularly in Africa, has been worrisome for researchers interested in maintaining a safe public and animal health system.[37,38]

The life of a pill is long and complicated. Processes and practices, tools and technicians, are involved in all of the steps that takes the pill from one part of the world to another, sometimes in barrels, sometimes in bottles, and sometimes in boxes. The beginning, middle, and end are separated by long processes, in time and space, that rely on each step working in synergy. Unfortunately our emphasis has only been focused on improving the quality of the whole life of a pill by looking only at the end of its manufacture.

3

An Age-Old Problem

The year 1857 changed the course of Indian history. The mutiny among the soldiers of Indian origin against the British East India Company ended the Mughal dynasty and with it the British East India Company. The unsuccessful mutiny forced the British Crown to have direct control over India, creating a new chapter in colonialism: the British Raj.

The new Raj required more civil servants coming to India from London, as well as a lot more soldiers whose loyalty was not questionable. The bureaucrats and the soldiers needed to be fit and healthy to work in the Indian environment and be prepared to face the challenges of the hot and humid climate. Among their supplies, they needed quinine bark, and lots of it, to fight malaria. Despite being in the market for quite some time, the quinine in India was costly and often adulterated. The quality of imported quinine from South America, coming through various middlemen, was highly suspect, and the last thing the government wanted was to have its soldiers fall sick in the new colony. Giving soldiers bad quinine would have done irreparable damage in moral and trust. A massive effort was started to create local plantations in India, that continued for as long as the British remained in India.

Quality quinine powder was also needed for a completely different purpose. Quinine was bitter, but a daily dose was required to stay healthy. A home brew, mixing quinine powder with sugar and soda, was created in India to make it more appealing to British soldiers and civil servants. In 1858, Erasmus Bond, a British entrepreneur, patented his version of the home brew, the tonic water for commercial sale.[1]

Gin and tonic, a favorite in the British Raj, required high-quality quinine.

The problem of substandard and counterfeit drugs is as old as the drugs themselves. Dioscorides, while classifying drugs for their therapeutic use and efficacy in the first century B.C., warned the users of adulterated drugs about their impact. But he did not stop at the warnings. He also felt that it was his duty to suggest ways to detect that fraud. His work in medicine looks at both the adulterations while at the same time contributing to methods in detecting fake, adulterated, or substandard drugs.[2]

From the earliest physicians up until recently, the practitioners of medicine and surgery wore multiple hats.[3] They were more than people who were just responsible for diagnosis, treatment, and management of the disease. Their work also spanned the area of drug discovery, experimentation with various potions and plants, and the analysis of therapeutic effects of their discovered medicines. It was their training and a requirement of their job to not only diagnose a disease and know what to prescribe, but also look for the best remedy.

These multitasking physicians played an important role in classifying drugs by not just their ingredients and their therapeutic efficacy, but also the right doses. As they described the potency, and prescribed a particular regimen of the drugs, they also warned against subtherapeutic doses and adulterated medicines.

While things may have changed in Western medicine, traditional practitioners of herbal medicine in many parts of Asia and Africa still continue to play the simultaneous roles of healers and pharmacists.[4] In cases the practitioner of medicine himself or herself does not create a therapy, he or she is still actively involved in training others who may be tasked to create therapeutic doses of the medicine in question.

Because the practice of prescribing drugs varied from one physician to the other, it is perhaps not surprising that herbal medicines, which have historically been given in the absence of regulation, have had a long history of being adulterated.[5] Cholera medicine in ancient times used to be a root called Valeriana Officinalis, which was routinely adulterated with red clay to deceive the unsuspecting consumer.[3]

However, probably the most famous incidents of fake medicine recorded started in the middle of the seventeenth century and continued until the earlier part of the twentieth century.[6] The drug being counterfeited was a blockbuster of its time, able to treat malaria (coming from the Italian words *mal* and *aria*, meaning bad air, as it was assumed to be related to areas that had foul air). This blockbuster drug was a bark from a tree in Peru, that grows to be about fifty to sixty-five feet and was known among the native population as *Quinquina* (the bark of barks).[7] The colonists called it Chinchona, naming it after the Countess of Chinchon, the wife of the Spanish Viceroy of Peru, who was treated by this bark in 1638. Just as the bark started gaining popularity, there was also strong resistance against its use. The original resistance against the bark was from the medical community, which labeled the use of the bark as magic and the Pope's secret plot to control lives of all those who opposed him. The medical community favored their own profitable practices that preferred bloodletting to achieve the elusive Galenic balance of blood and bile. It was only when royalty, Charles II and Louis XIV and their families, were saved by the bark that the medical community accepted this as a potent medicine. Because of the Jesuits bringing the bark to Europe, it was commonly called Jesuits' bark instead of Chinchona bark.

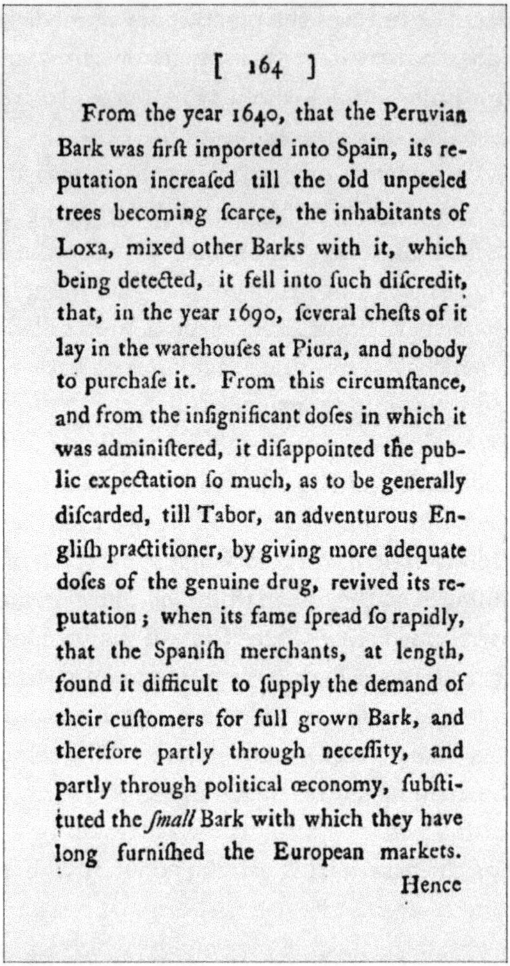

[164]

From the year 1640, that the Peruvian Bark was firſt imported into Spain, its reputation increaſed till the old unpeeled trees becoming ſcarçe, the inhabitants of Loxa, mixed other Barks with it, which being detected, it fell into ſuch diſcredit, that, in the year 1690, ſeveral cheſts of it lay in the warehouſes at Piura, and nobody to purchaſe it. From this circumſtance, and from the inſignificant doſes in which it was adminiſtered, it diſappointed the public expectation ſo much, as to be generally diſcarded, till Tabor, an adventurous Engliſh practitioner, by giving more adequate doſes of the genuine drug, revived its reputation; when its fame ſpread ſo rapidly, that the Spaniſh merchants, at length, found it difficult to ſupply the demand of their cuſtomers for full grown Bark, and therefore partly through neceſſity, and partly through political œconomy, ſubſtituted the *ſmall* Bark with which they have long furniſhed the European markets.

Hence

FIGURE 3.1.

An account of a fake Chinchona bark business prospering in Europe in the seventeenth century. As the businessmen were unable to meet the demand, the fake bark was being used as a substitute for an authentic Peruvian (Chinchona) bark. Reproduced with permission of the Wellcome Trust Library of the History of Medicine, London, UK.

The bark was a highly valuable commodity and soon started to make an impact outside Europe. It appeared in the courts of Peking, China, and Kyoto, Japan. There, they were used to cure the ailing royalty and the high-powered courtiers. These regions were far from the original discovery lands of Latin America and the markets of Europe. The demand for the bark was so substantial, that in the nineteenth century new plantations appeared in America and the first cinchona plantations outside South America were created.[7]

Malaria, and efforts to effectively treat it, affected individuals and armies, and it influenced colonial efforts and policy of engagement with the local tribes.[8] The cases of counterfeiting and adulteration of Peruvian Chinchona bark, started soon after the discovery of the potency, and continued until the mid-1940s.

From royalty to the colonists, there was a high demand for a potent antimalarial. Driven by this high demand, barks that resembled the Jesuit bark started appearing in the market, which were misleading naïve consumers and were useless in treating patients. But the issue was not just of efficacy. It was also of quality control. The issue of genuine bark, and not a substandard or a knock-off, was critical.

In the seventeenth and the eighteenth centuries, the main strategy against counterfeiting was better storage and trust on the supply side. The Jesuit colleges and Jesuit pharmacies were more trustworthy, and hence their storage of the bark was considered particularly valuable and efficacious.

But there was only so much that could be stored in a Jesuit pharmacy or college, and the Jesuits were not everywhere to store the real bark. There was also religious resentment against the Jesuits. Further, in the marketplace were many who wanted to profit by adulteration or by simply selling a counterfeit. Regulatory mechanisms that could check quality and enforce laws were often lacking.

In the absence of regulation and testing, the presence of fake bark that did not work started worrying consumers. This decline in confidence at the consumer level had a devastating effect on the Chinchona bark (the drug derived from this bark later came to be known as quinine) market.[9,10]

Despite the impact on both the consumers and the markets, the issue of substandard and fake quinine continued for some time. The reports were coming from all around, affecting confidence and even the morale of those who were in the tropical climate.

Something had to be done.

Lancet took the lead in creating awareness. In an issue published on April 29, 1829, the journal said "the adulteration of quinine is carried to a greater extent than is generally supposed, while the necessity of having it genuine is most important. The high price it obtains, renders it a source of successful imposition to sellers, and of corresponding disappointment to consumers; and this is most probably the cause of the various degrees of benefit with which it has been used in the same complaint. The adulterations most frequently used, are, a peculiarly fine preparation of crystals of spermaceti, starch, and the gentianae, a preparation partaking in a great degree (when carefully made) of the appearance of quinine, but cheaper and quite useless."[11]

This letter in *Lancet* prompted responses from the public about how to counter the problem. Numerous people offered their own homegrown solutions and remedies to address the issue. Recipes, tricks, and strategies were sent to *Lancet* to distinguish a good quinine from a fake one.

In his letter to the editor of *Lancet* in 1838, the surgeon Charles Croft wrote, "You will confer a favour on the profession by giving publicity to the following test for discovering salicine, with which the sulphate of quinine is adulterated very largely, from its physical character, being very similar to that important alkaloid. In a letter received from M. Pelletier, of Paris, he directs twenty drops of the pure and concentrated sulphuric acid to be poured on twenty grains of the suspected quinine, when the solution will present a most beautiful crimson colour, more or less intense, according to the quantity of salicine present. The adulteration of one part of salicine with ninety-nine of quinine is, by these means, easily discovered."[12]

The public awareness campaign, though an important step, was far from sufficient. The problem did not disappear, and in parts of the world, it continued unabated and even increased in its scope. The problems in the British colonies, particularly those in the tropics and hotter climates, were particularly troubling.

In the twentieth century, the problem of adulterated quinine had taken a distinctly modern form. It was no longer a bark to be chewed, or powder that looked as if it had been made from the Jesuits' bark. It was now a tablet, that appeared, in shape and size, like a genuine one. But in fact it had little to no active ingredient. In 1932, the editor of an Indian medical gazette noted that "we can think of no more despicable act than selling to a malaria-stricken peasant as 'quinine' a tablet containing nothing but chalk or some such inactive substance. . . . If these are a fair sample of quinine tablets on the market, it is obvious that the most serious adulteration is going on within the country."[13]

While it may seem that modern technology, better regulation, and awareness in general may have resolved the problem, at least for quinine, in reality it has continued to persist with immense stubbornness. The problem of poor-quality, substandard, or outright counterfeit quinine has been widely reported in scientific and general audience literature; it has been very difficult to resolve it completely. Even in the so-called developed or high-income countries, the problem does not seem to go away. Between 1969 and 1992, the US Food and Drug Administration (FDA) received 157 reports of health problems related to quinine use, including 23 that resulted in death. The continued challenges with quinine resulted in the FDA ordering to stop the use of quinine for nighttime leg cramps. Eventually, The FDA also took away the over-the-counter status of quinine.[14]

Quinine and its derivatives (particularly quinine sulfate) continue to be sold in subtherapeutic (i.e. a lower dose than what it is supposed to be), substandard, and counterfeit forms around the world. As recently as 2014, the Ghana Food and Drug Authority warned the public about fake quinine sulfate that was produced by an Austrian company. It was regularly sold and widely available in the local market, not just in Accra but in other parts of the country as well. In its statement the Ghana FDA said "results of laboratory analysis conducted on the product at the FDA Laboratory showed that the Quinine Sulphate Tablet contains no active pharmaceutical ingredient making it a complete counterfeit/fake medicine" and issued a stern warning that "the general public is entreated not to patronize the said product and to report to the FDA anyone found selling them."[15]

Quinine is not the only drug that has had a tainted past, in spite of large-scale efforts by citizens, organizations, and even governments to curb the commerce of its fake and poor-quality substitutes. A sweet raspberry-smelling syrup has also had a devastating, albeit a more recent, history.

In 1937, Archie Calhoun was a physician in Mt. Olive, Mississippi, practicing in his clinic on the second-floor office at Powell Pharmacy on Main Street. Mt. Olive is part of Covington County, located in the southern part of the state. In 1930, the census showed about 15,000 people living in Covington County and just under a thousand in Mt. Olive. Because Covington County was relatively small, the patients of Dr. Calhoun used to come from all over the county. Many of his patients were his acquaintances and even friends.

Dr. Calhoun would see patients with a variety of ailments, including common problems like strep throat. To cure strep throat, one of the most potent medicines of the time was Sulfanilamide, a drug with remarkable curative properties. The drug was first prepared by an Austrian chemist, Paul Josef Jakob Gelmo, in 1908 and was patented in 1909.[16] The drug was a potent antibacterial and used widely by the military, particularly by the Allied powers, in the second world war to combat infection in the war zones.[17]

The drug was used initially in a powder or a tablet form. However, there was an increasing demand for the drug, in many southern states in the United States, in a liquid form. Responding to this demand, Harold Cole Watkins, who was a lead chemist and pharmacist at S. E. Massengill, in Bristol, Tennessee, discovered that the drug dissolved in diethylene glycol (DEG), a compound highly toxic to animals and humans. The company called this new formulation "Elixir Sulfanamide." To make it more appealing, raspberry flavoring was added to the solution.[18]

The company did test the drug for flavor, appearance, and smell, but never tested it for toxicity. Toxicity tests were not required by the law at the time. A total of 633 shipments of the elixir mixture were sent all over the country. Dr. Calhoun was among those physicians who prescribed this liquid drug to his patients. But unlike the previous instances of using the powder form of Sulfanamide, which was widely used, the elixir's reaction was unexpected, complicated, painful, and fatal. What was supposed to cure strep throat became a deadly potion. This was heartbreaking for Dr. Calhoun who was personally attached to his patients.

On October 22, 1937, Dr. Calhoun wrote in a letter, "But to realize that six human beings, all of them my patients, one of them my best friend, are dead because they took medicine that I prescribed for them innocently, and to realize that that medicine which I had used for years in such cases suddenly had become a deadly poison in its newest and most modern form, as recommended by a great and reputable pharmaceutical firm in Tennessee: well, that realization has given me such days and nights of mental and spiritual agony as I did not believe a human being could undergo and survive. I have known hours when death for me would be a welcome relief from this agony."[18]

Archie Calhoun was not the only one who saw this tragedy. The liquid form of the drug was sent to many other clinics and pharmacies and the reports of death started to quickly pour in from Oklahoma, New York, Virginia, and even California. The organ failure that resulted from the consumption of the elixir was extremely painful for the patient, and torturous for the caregivers to watch. The problem started to gain national attention. Letters were written to President Roosevelt, including by a woman who described the death of her child as "the first time I ever had occasion to call in a doctor for [Joan] and she was given Elixir of Sulfanilamide. All that is left to us is the caring for her little grave. Even the memory of her is mixed with sorrow for we can see her little body tossing to and fro and hear that little voice screaming with pain and it seems as though it would drive me insane."[18]

All in all, more than one hundred people in fifteen states, across the nation, died. The chief chemist, Harold Watkins, when pressed to admit guilt, argued that "we have been supplying a legitimate professional demand and not once could have foreseen the unlooked-for results. My chemists and I deeply regret the fatal results, but there was no error in the manufacture of the product. I do not feel there was any responsibility on our part."[19] Eventually, Harold Watkins took his own life.[20]

The national outcry, a reflection of a sense of panic and loss, along with what was required of pharmaceutical companies, led to the passage of the 1938 Food, Drug, and Cosmetic Act. It was a turning point in the history of drug regulation in the United States and the creation of the modern FDA.[18,21]

Yet, despite the creation of the FDA and the significance of diethylene glycol (DEG) in strengthening regulation of drugs, the compound has continued to surface as a killer of children and adults around the world, even in the twenty-first century.[22] In Mumbai (formerly known as Bombay) in 1986 twenty-one patients died of kidney failure when glycerine was contaminated with DEG.[23] In 1998, in Delhi, thirty-three children died between April 1st and June 9th, when the solvent syrup contained DEG. The children were aged between two months and six years and admitted to two hospitals in Delhi.[24] The children died of acute renal failure, a situation similar to what was seen nearly sixty years before in the United States. The cause of death was a cough syrup, locally made, that had DEG.

One of the worst disasters associated with DEG was in Bangladesh in 1992 where "countless children" died as a result of contamination in a commonly used fever medicine, Paracetamol.[25] The number of those who were affected is still debated, with some estimates suggesting that those who were affected were in the thousands.[26] In response to the "epidemic" of renal failure, the government banned the sale of Paracetamol, which was shown to contain DEG.

In Haiti, in 1996, eighty-eight children died because of DEG contamination of acetaminophen (the key ingredient in Tylenol and Panadol) syrup.[27] Children were aged between three months and thirteen years with nearly 85 percent of them under five. The Centers for Disease Control and Prevention (CDC) report of the tragedy in Haiti paints a heartbreaking picture. It says "most cases were characterized by a nonspecific febrile prodromal illness followed within 2 weeks by anuric renal failure, pancreatitis, hepatitis, and neurologic dysfunction progressing to coma. Ten children were transferred to medical centers in the United

States for intensive care and dialysis; nine are still living. Of the 76 children who remained in Haiti, only one is known to have survived. Histopathology of kidney tissue from four patients indicated acute tubular necrosis with regeneration consistent with a toxic exposure."[28]

In 2006, families in Panama reported that nearly four hundred people died by taking a particular brand of cold medicine, which was later found to have DEG.[29] Most recently, in 2008, Nigeria reported deaths of young children given teething medicine (My Pikin) that contained DEG.[30] The *New York Times* reported that at least eighty-four deaths of children were due to DEG that resulted in acute renal failure. Nigeria had suffered a similar mass poisoning tragedy in 1990.[31]

The CDC estimates that in the last seventy years, despite awareness and highlighting of the issues, stories in print media and research studies in scientific literature, better laws and governance, and even the formation of new national regulatory bodies, at least twelve mass poisonings have occurred around the world.[28]

Roger Bate in his book *Phake* points out that there have been other poisonings and DEG contamination in consumer products across the globe.[20] DEG, due to its solubility, cheap production, and similarity in chemical structure, has continued to appear in consumer products. It has been found to be present not only in teething syrups and antifever drugs, but also in consumer products such as toothpastes.

In most cases associated with DEG, it has been used as a solvent to dissolve one of the key active ingredients in the drug and not as an active ingredient itself. The supportive role of DEG makes the problem of detection and regulation more challenging for the concerned authorities.

Times where the prescription would be chewing on a particular bark, or a herbal concoction where all the ingredients were visible, have long passed. The modern drug, even in its tiny size, is a composition of dozens of compounds and materials, some of which are used only in trace amounts, yet could be deadly if used in the wrong concentration or substituted by similar but toxic analogues. It is no longer simply about the drug itself, which can be compromised or completely absent from the pill, but also about other ingredients that are being used increasingly to make the drug more soluble, more palatable, or simply better looking. That complexity is only increasing for regulators posing newer and more complex challenges.

This begs the question about technological advancement and governance of quality, and how we manage an age-old problem. Is it getting worse, not just because there are more drugs on the market, but also because the drugs have so many parts and components? Is it that the modern drug, with all its immense power and marvelous potency, specificity, and impact, makes the problem of testing and regulation harder? Has the complexity of modern drug manufacture made it easier to cheat? Will regulation always be behind, and will we perpetually wait to create better regulation until something goes wrong?

4

Of Mice and Cats

It had been another long day at work for Samuel.* His task this month, like last month, was to collect samples from various parts of the city, ranging from the fever-reducing and widely available Paracetamol to antihypertensive drugs. He wanted to collect as many samples as he could from about a dozen pharmacies and then log them and send them off for testing. The testing lab wasn't far from his office, but in some ways it was really far. With the amount of time it took for the drugs to be tested, it was as if they were sent to another planet.

Today, like last month, when he came to the market in his old car, he saw a familiar scene. Standing on the street corner were young men, sitting idly. As his car stopped, the boys saw him and immediately got up. One started talking on the phone and another started walking briskly. Pretty soon, all the pharmacies, lined on the street, had their shutters down. The pharmacy workers had been told that an inspector was coming.

Once again, he came back empty-handed back to his office.

When I saw Samuel in Nairobi, he talked about the game of cat and mouse that was being played constantly. One of his tasks as a drug inspector in the city was to collect samples, but it was not easy. On some days, he could go early and surprise

* The name of the drug inspector in Nairobi is changed to protect his identity.

the pharmacies, but on others, the pharmacists were tipped off and they quickly closed the shop and would either leave from the back door or keep the shutters down and wait until he was gone. He did not have any authority to force them to give samples, and the agency did not have the resources to employ secret buyers on a regular basis. Sometimes he won, and sometimes the pharmacists won. For as long as he could remember, it had always been like this.

Unknown to him, Samuel was and remains a part of a long tradition of inspection, and he was not the first inspector in the history of drug regulation who had been frustrated by the difficulties of his job.

While the problem of substandard and counterfeit drugs is hundreds, perhaps thousands, of years old, so is the effort of individuals, communities, and governing bodies to address this problem.[1] For millennia, the game of cat and mouse has been played on all stages. Most people have heard of the Food and Drug Administration (FDA) in the United States or the Medicines and Healthcare Regulatory Authority (MHRA) in the United Kingdom, but these agencies and organizations, while often in the news, are relatively recent institutions. They, like many other regulatory bodies over the course of history, have been created to respond to immediate threats and crises. The story of regulation dates back nearly three thousand years.

There is strong evidence both from the Egyptians and the Greeks about the desire to control medicine quality. This desire has been an integral part of the medical profession from ancient times. The Hippocratic Oath, for example, reminds the physicians against prescribing a poison or giving of "pharmacon oudeni," loosely translated as "I will give no deadly drug."[2] Similarly the inscription on the Athenian Acropolis from the fourth century commemorates Evenor the Physician who was considered a drug inspector and entrusted with quality checks.[1] The Greeks, through their various texts, refer to the Egyptians, whom they considered to be particularly concerned and serious about drug quality. Homer, in his book *The Odyssey* remarks about the Egyptians saying that "there the earth, the giver of grain, bears the greatest store of drugs, many that are healing when mixed, and many that are baneful: there every man is a physician, wise above the human kind, for the yare the race of Paeeon."[3,4]

Diodorus Siculus, a famous Greek historian who lived in the first century B.C., was also concerned with drug quality and suggested harsh penalties for those who fail to prescribe quality remedies. He argued that "the physicians draw their support from public funds and administer their treatments in accordance with a written law which was composed in ancient times by many famous physicians. If they follow the rules of this law, as they read them in the sacred book and yet are unable to save their patient, they are absolved from any charge and go unpunished; but if they go contrary to the law's prescription in any respect, they must submit to a trial with death as the penalty, the lawgiver holding that few physicians would ever show themselves wiser than the mode of treatment which had been closely followed for a long period and had been originally prescribed by the ablest practitioners."[5]

Not just historical documents or literary masterpieces argue for quality control; it has also been discussed in matters of faith and religion. The history of regulation, and encouragement to test quality, as well as admonition of those who do not develop or prescribe the highest-quality medicines is weaved within the religious literature. The Old Testament in Ecclesiastes 10:1 says, "Dead flies cause the ointment of the apothecary to send forth a stinking savour."[6]

The interactions between religion and the institutions of the state that framed its structure, governance, and organization on religion are perhaps most evident in the early Muslim government and the office of the "*Hisba*."[7] The office created in the early ninth century, and lasted several hundred years, was the custodian of ethics, law, and regulations. The fundamental mandate of the office was accountability and enforcement of a certain code. The early incarnation of the office of Hisba focused largely on moral ethics. However, as the Muslim empire expanded and new governing structures were created overtime, the office of Hisba grew to accommodate other areas of social interactions, including between patients and caregivers. The person in charge of the office was called "Muhtasib"—a term that can be loosely defined as the ombudsperson. The term Muhtasib is still used in many parts of the Muslim world, and has become part of the local vernacular, including in countries, such as Pakistan, where Arabic is not spoken.

In different parts of the Muslim world, the role of Muhtasib varied tremendously. In some places, the Muhtasib was a judge, someone with an authority to preside over a trial and hand out punishments; in other parts the Muhtasib was someone who encouraged good moral practices for social cohesion and fairness within the society, and yet in others the Muhtasib was an inspector, someone was in charge of inspecting grain, enforcing weights and measures, and testing the quality of various remedies. Nonetheless, in all of these circumstances, the position of the Muhtasib was an important position in the society, representing both prestige and authority. The role of Muhtasib was also tied strongly to the religious edict, based on both the injunctions in the Quran and the teachings of the Prophet Muhammad.[7]

The writings of various Muslim scholars and jurists like al-Mawardi[8] and al-Ghazali[9] cover various roles and responsibilities of the Muhtasib. They, along with other leading intellectuals and jurists of their time, paid particular attention to the need for inspection of pharmacies and the manufacture of bread and perfumes. As pharmacies were the location where the drugs were not just sold, but also manufactured, there are specific instructions in the texts of Muslim jurists for the Muhtasib on how to conduct themselves and what to inspect.[7] The work of Ibn al-Ukhuwwa focuses particularly on the role of the Hisba on the manufacture of syrups and preparation of drugs by pharmacists. He states that a physician should be able to understand the organs and their ailments and "useful drugs for them, ersatz drugs for those not available and methods of their extraction, and their manner of treatment so as to steer in a quantitative balance between illness and remedy and to counter the illness with the drug's qualitative properties. Those who can not do this should not be entrusted with treatment of the sick."[7,10]

Ibn al-Ukhuwwa also comments that in case the death of the patient was caused by negligence of the physician who makes up the prescription, the Muhtasib should tell the relatives of the deceased to "demand blood money from the physician for your kin for he killed him by his lack of art and his inadequacy."[11] Ibn al-Ukhuwwa goes back to the Hippocratic Oath and reminds the Muhtasib that all physicians should adhere to the oath and that physicians shall not "prescribe a poison for a patient, they will not describe poisons to the people, that they will not give a woman a drug which aborts the embryo nor a man to avoid conception." Ibn al-Ukhuwwa also makes a strong case for regulation of drug pricing and the ethics of pricing in the marketplace.

Another important contributor to the discourse on drug quality was Hunain ibn Ishaq. He was a prominent Arab physician and scientist in ninth-century Mesopotamia, focusing particularly on ophthalmology and ophthalmic surgery.[12] In describing the roles of the Muhtasib, Ibn Ishaq argues that Muhtasib must ensure that the ophthalmologist must know how to compound the collyria and mix the drugs.[13,14]

The office of Hisba was not just about creating a framework for regulation, but also about exposing the charlatans and frauds. The code of Hisba categorically admonished frauds for their trickery and suggested harsh punishments for their conduct. On a practical side the office of the Hisba was instructed to carry out stealth operations and the staff of the office was encouraged to go after hours such that the "syrups and drugs may be inspected anytime without warning, including after their shops are closed for the night."[15]

Similar to the modern pharmacopeial approach, the office of Hisba was also a central body in creating standards for making quality drugs. Standards were issued for making good syrups with a list of which ingredients to use and which ones to avoid. For example, one such instruction suggests that for syrups, only good Egyptian spotless sugar should be used and not honey, and even the sugar and the fruit juice should have the right proportions.[16] The office of Hisba had syrups that were registered, along with their ingredients, and could be inspected for both their composition and their quality.

The evolution of the office of the Hisba presented a structural and organizational departure from earlier approaches in drug-quality regulation. The approaches by the Egyptians and the Greeks focused largely on instructing physicians to reflect on their responsibilities and conduct their work with the highest moral and ethical standards. The office of the Hisba, which initially focused on moral conduct and was an office where complaints could be lodged, evolved from this guardian of moral ethics into an organization that also conducted tests and inspections, and it eventually started to issue guidelines about what should be in a drug and how often to sample it. The long and detailed discussion on adulteration and fraud in manufacturing, the causes, the obvious signs, and how to address this in society reflected this change in the role of governance. The creation of the Hisba moved the responsibility, although not completely, from the physician to a regulatory body to maintain quality.[11]

That said, one of the big challenges in the office of the Hisba and other similar prior efforts was the lack of appropriate tools to inspect drug quality. The best available tool was visual inspection, which came with experience. This was supplemented with smell, and in some cases, taste. Occasionally, weights and measures were also used, but they were not necessarily useful in determining quality. In the absence of precise tools, the consequence of a bad or adulterated drug was evidence based on worsening condition of the patient (or his or her death). The tragedy of the loss of life led to the punishment of the drug maker, but did little to improve general drug-quality regulation.

In Europe, the origin of drug regulation came from a different source. It was not the church, the police, or the office of enforcers. Instead a major part in drug regulation was played by the new medical schools appearing across Europe and the curricula of instruction at these institutions.[17] In particular in the tenth century the school of medicine in Salerno (modern southwest Italy) was enjoying widespread acclaim for its novel approach to education and the practice of medicine.[18] The medical school also trained its students to become inspectors of drug shops. The goal of creating new inspectors was two-fold: it was for public health and safety, and for protection against contagion.[17]

This new medical school was well ahead of its time. It argued and practiced a new model of training, where the roles of apothecary and the physician were separated, something that had not been done up until that time. Additionally, there was an ethical component of training as well. To avoid any future conflicts of interest and to protect public health, medical students, who underwent an eight-year course, were instructed not to keep an apothecary shop or even get into any business relationship with any pharmacy. Similarly, during the same period, the apothecaries were trained and required to maintain a strict standard, by the use of weights and measures, and instructors for apothecaries were appointed. The apothecaries were also trained and instructed to keep only a professional relationship with the physicians and healers.

Despite the difference in origin, there were also similarities between the Hisba approach and the regulatory practices in Europe. For example, the penalties for malpractice were harsh and nearly identical. Selling or dealing in poisons were strictly forbidden, with a law stating that anyone found dealing in that craft will be hanged.[17] The separation of craft of physicians and apothecaries, started in Salerno, was used subsequently in other parts of Europe, in particular Basle (also written as Basel) where the apothecary had to present himself in front of the judges.[19] There was a test that the apothecary had to pass; only then the apothecary could take an oath promising to provide the physicians with drugs of the highest quality.[15]

The approach in Europe that had started in the tenth century was more at the grassroots level and aimed to empower physicians and pharmacists through training and awareness, and an understanding of how to conduct themselves professionally. However, this was not the only approach to regulate drug quality. There were also top-down efforts to create laws and frameworks to ensure quality. From a state-control standpoint, the major development in Europe came in 1140 with Roger II, the King of Two Sicilies, promulgating the first known law of drug regulation in Europe, stating that "whosoever will henceforth practice medicine, let him present himself to our officials and judges to be examined by them but if he presume of his own temerity, let him be imprisoned and all his goods sold by auction."[20]

This tradition of strengthening laws to regulate medicines continued in the royal family. Roger's grandson, Frederick II, who ascended the throne when he was very young, issued additional laws on this matter during 1231–1241. These laws were similar in spirit to his grandfather's but much broader in scope. They were also enforced over a much larger geographic region. The laws in Frederick's reign were enforced not only in Naples and Sicily but also as far north as Germany and France.[17]

These promulgations and laws, which started in the Holy Roman Empire, started to create a culture that emphasized quality and purity of substances. This culture created a subsequent demand for the official control of drug quality. While there were some efforts at the level of individuals and some training of doctors, a societal-level culture and a demand did not exist before the efforts of Roger II and his grandson, Frederick II.

As the culture of drug quality got official backing, new traditions of elaborate practices, to celebrate preparation and certification of drugs of high value and efficacy, started to spring up across Europe. This was apparent in the celebratory festivals in Venice, France, and Germany in the thirteenth century. In these places, the preparation and certification of quality for Theriac, an ointment that was used as an antidote to venomous bites, was an event with pomp and ceremony that would have the feel of a town fair, with manufacturing and preparation in public.[21] The celebratory practice of certification and even manufacture continued in Europe for several centuries. John Evelyn, a famous English diarist, in his diary entry of March 23, 1646, writes that "the making of extraordinary ceremonie whereof I had been curious to observe for tis extremely pompous and worth seeing."[22] Theriac in Venice was certified with an official seal of the republic and often stored in ornate porcelain jars. The central place of Theriac and its certification continued in official documents and was a part of the pharmacopeias of Germany and France even in the nineteenth century.[23]

Britain's history of regulation was neither from the medical school nor from the king or the office of an ombudsman. Instead, it started with the pepperers and grocers with the Ordinance of the Gild of Pepperers of the Soper Lane in 1316, which is probably the earliest written code of drug-quality control in England.[24] The Soper Lane, a street that no longer exists in London, was the location where grocers, curriers, and pepperers operated for centuries, until the street was badly damaged in the great fire of 1666.[25] The pepperers took over the distribution of drugs, spices, and other mixtures sometime during the early part of the twelfth century.[26] The ordinance of the Gild of Pepperers, which focused on maintaining the quality and standard of spices, included apothecaries as well, because mixing was involved, and medicines were based largely on herbal formulations that were grinded and mixed to form syrups and powders.

Prior to the Renaissance in Europe, while there was strong evidence of drug inspection and individual cities such as Venice that had established rules and regulations for inspections and code of conduct for inspectors, there was little separation between pepperers, grocers, and apothecaries as separate craftsmen. For several hundred years, even after the Ordinance of the Gild of Pepperers, there was little recognition for the separate training, roles, and regulations needed for apothecaries that was distinct from other herbalists, spicers, and grocers. There were clear distinctions and a code of conduct for apothecaries from physicians and healers, but the line between various businesses that relied on mixing, milling, and grinding was blurry. This blurry line had implications for inspectors, who were in charge of inspecting spices and peppers, as well as herbal remedies that could make a difference between life and death.[27]

Two additional features of the Middle Ages were particularly unique in terms of drug inspection and the regulations that governed them. First, the drugs that were checked (and there is little evidence to suggest how many or how frequently were they checked) were largely checked for adulteration and for the right weight. A strict code of weights and measures was used to test whether a drug was qualified to be supplied, sold, or administered and at what price. This emphasis on weights and measures had less to do with medicine and more to do with the history of regulation that was borrowed from regulating grain, spices, and other herbs. In spices, grains, and herbs, falsification through wetting was common, and hence the practice of weight measurement made its way to drug-quality inspection as well.

Second, and perhaps more important, the efficacy of the drug was never in question. The testing of the drug or the syrup was based solely on its ingredients and its weight, never based on its ability (or inability) to cure a particular ailment. Similarly, issues associated with whether a particular drug would lead to a negative or an adverse reaction was not within the realm of quality control or testing of any kind. The practice of regulation was focused more on the commerce and pricing, and the term "drug regulation" of that time reflects more the checking of physical attributes of the drug than its clinical efficacy or what we may today call bioavailability.[17]

Like so many practices that changed with the Renaissance in Europe, the practice of drug regulation was also transformed. The Renaissance changed the practice of drug regulation in three important ways. The first was the creation of organized institutions, such as the colleges of physicians and surgeons and the company of physicians, that started to appear in various parts of Europe. The creation of these institutions provided a concrete structure to the role of physicians in the society and also gave them the authority to appoint apothecary inspectors. In 1518, Thomas Lincaire obtained the charter by Henry VIII to establish the "Company of Physicians," which later become the Royal College of Physicians of London in 1551.[28] In 1540, a statute on the control of drugs was passed in Britain. This landmark statute changed the world of pharmacopeia regulation in a profound way, as it now gave the physicians the right to search the shops of apothecaries.[27] Additionally, it also gave the physicians the authority that during a search if they found "drugs that were defective corrupted and not meet nor convenient to be ministered in any medicines for the health of man's body, then the searchers were to call for the Warden of the Apothecaries and the defective wares were to be burnt or otherwise destroyed."[29] This was further strengthened by Henry VIII when he stated that "in the Kings court . . . no water of law, essoin [excuse] or protection shall be alloweth."[1]

The first such visit, as a consequence of this law, which was a combination of sample collection and inspection of drug quality, took place on January 25, 1542.[28] The practice of inspection by the members appointed by the College of Physicians continued for over three hundred years until the last visit on November 2, 1858.[1] This right of the visitation was withdrawn with the medicines act of 1875.

The second major change during the Renaissance was the organization within the apothecaries. Like the physicians, they also formed independent and organized structures among themselves. King James, who was James VI of Scotland from 1567 and became James I in 1603, upon unification of the Scottish and English crowns, established the society of apothecaries in 1617. This act separated the apothecaries from the grocers. This was a major victory for the apothecaries as they considered their craft to be fundamentally different than the grocers, and they wanted to come out of the shadows of grocers and pepperers.

James I established them as "The Worshipful Society of the Art and Mistery of Apothecaries."[29]

Things were not quite as straightforward, however. Nearly seventy-five years before the decree by James I, King Henry VIII had given exclusive rights to the physicians to appoint inspectors to test the quality of products sold by the apothecaries. The practice of apothecaries was therefore at the mercy of inspection by those appointed by the physicians, and the two were often not on the same page. This led to strong tension between the two groups. To diffuse this tension and develop a working relationship, a group of apothecaries and physicians met and agreed that the apothecaries will not supply cathartics, vomits, or sudorifics without the knowledge or consultation of the physicians.[1] This also led to the agreement that a new book called "London Antidotary" was to be published by the College of Physicians for the guidance of apothecaries so that they could stock their shops with appropriately prepared medicines.[30] A major disagreement between these newly recognized apothecaries and the physicians was whether the apothecaries could sell the drugs directly to the public without an explicit permission or prescription. The apothecaries wanted to contest any such move by the physicians, as it would affect their overall business. Not being able to sell directly to the public also meant that their trade and craft would no longer be controlled by them and would always depend on the judgment and whim of the physicians.[29] The physicians, on the other hand, were adamant that such an order was needed for public safety and control of increasing levels of corruption among the apothecaries. The apothecaries created strong opposition to the pressure from the physicians and pointed to the grocers—who ironically were the group that the apothecaries had distinguished themselves from some time ago—as a group that was able to sell products directly. The apothecaries despite their best efforts lost the political and the legal battle, and the order requiring apothecaries to get an explicit approval or permission from physicians was passed in 1632 and was added to the "Freeman's Oath of Apothecaries."[1,31]

The third major development of the Renaissance was the changing face of drug manufacture and the control needed for that. This was a time when medicine making was being streamlined, in part due to interactions between scholars and physicians from various parts of Europe (and beyond), and in part due to the growth in volume of available drugs that needed standardized procedures for protecting drug quality and manufacture. It was during this period, that the first pharmacopeias (from the Greek words *Pharamakon*, meaning drug, and *poiia*, meaning making) started to appear on the regulatory horizon. The purpose of a pharmacopeia during this period was not only to standardize ingredients but also to provide instructions for the compounding of medicines. At the same time, the pharmacopeia document was also supposed to be the official guide for manufacture for everyone, therefore bringing with it the government seal of approval. This new document was the bible of regulation, and was to serve as a benchmark for drug inspection and to determine quality against a set of standards.

The modern pharmacopeias trace their roots to Florence, Italy. The first official formulary (a list of ingredients and instructions in compounding) was issued in 1498 by the Florentine guild of physicians and pharmacists. This was the obligatory guide for the apothecaries of Florence.[1] A more comprehensive pharmacopeia, that went beyond the formulary of Florence, was by Valerius Cordus (1515–1544), whose posthumous edition of "dispensatorium pharamcopolarum" required all apothecaries in Nuremberg to make all their preparations in accordance with Cordus's book.[32] With the growth and recognition of the importance of streamlined procedures detailed in the pharmacopeia, other cities, states, and political units soon followed with issuance of their own pharmacopeias.

The London pharmacopeia was published in 1618, with the first official edition appearing on December 7.[33] While in Latin, the preface of this pharmacopeia is telling in many ways. First, the preface is written for the "candido lector" or the friendly reader, not just for a physician or an inspector alone. Second, it talks about the fact that there were many books available on the topic, from various parts of the world. This was a clear way to suggest that there was a need for a standard, a universal code, and this particular pharmacopeia was filling that gap. This was an implicit way to establish authority.[34] The preface of the first edition gives some clues toward this goal by saying that "we do not instruct, but impose and enforce one and only law, and one method of compounding, from which we would allow no deviation whatsoever, not even that of a finger breadth."[33]

Later on, the preface asserts its unique position further by saying that "for the future, we have removed all such power arbitrary decision on the part of the compounder. We have adopted uniform measures and have determined upon a certain dose which the pharmacist shall neither increase, nor decrease."[33] Finally, the pharmacopeia's introductory text is also a window into the state of affairs at the time. It says that this "book will counteract, namely the very noxious fraud or deceit of those people who are allowed to sell the most filthy concoctions, for the sake of profit and instigated by sordid avarice."[33]

The London pharmacopeia was another turning point in the history of drug regulation. Issued first in 1618, it was updated and published at various intervals for nearly two hundred and fifty years. The various editions built upon the previous one, added new formations, and reflected changes in industry and practice of the time. The last edition of the London pharmacopeia was the tenth version in 1851. In that year, pharmacopeias from Edinburgh and Dublin, which had been publishing their own versions, were combined within the British pharmacopeia.[35] The British pharmacopeia has been published regularly since the unification with the Scottish and Irish pharmacopeias.

The combination of new demand for more and standardized drugs, the evolution of apothecaries as separate craftsmen from grocers, and the institutionalization of practitioners of medicines laid the foundations for drug regulation in Europe. This subsequently formed the basis of modern drug-regulatory authorities around the world, including the US pharmacopeia[36] and the FDA.[37]

The roots of modern drug safety, at least in the countries that saw major changes in their economies and the quality of life of its people, took root in the nineteenth and the twentieth centuries. Standing on the shoulders of the Renaissance, several major changes and turning points led to the concept of modern drug regulation and drug safety.

The first contributing factor was the birth of modern pharmaceutical sciences.[38] As the pharmaceutical sciences developed and matured, and found its footing on modern chemistry, it led to the birth of globally operating pharmaceutical companies. Similar to the factors that led to the formation of regional and then national pharmacopeias in Europe, it was development of new medicines that led to a major shift in drug regulation.

While laws such as the Apothecaries Act of 1815 had strengthened the society of apothecaries in England, it was the later part of the nineteenth century that saw a major transformation. With advances in chemistry, particularly in synthetic organic and inorganic chemistry, the development of new industrial processes such as the Haber process (which enabled the production of synthetic fertilizer) and the growing demand for industrial chemicals, a new pharmaceutical industry started to develop.[38] Major discoveries in chemical formulations and the shift from natural processes that were long, complicated, and inefficient to lab-based synthetic processes made it possible for companies to replicate their success. Making chemicals, including drugs, was now more systematic and reproducible. Drug manufacture was transformed from the domain of a craft to the domain of science.

New pharmaceutical companies started appearing on the market, such as Pfizer in 1850, Ciba-Geigy (the precursor of Novartis) in 1859, Lilly in 1876, Abbott in 1888, and Merck in 1890, during this period. These pharmaceutical companies utilized scientific literature in drug design and collaborated across borders. Drug design was no longer based only on experience, but on rigorous scientific principles of chemistry. The pharmaceutical companies also benefited from the development of other chemical industries, most notably the industry of synthetic chemical dyes, that transformed the fabric industry in Europe and beyond.[39,40] This development of a scientific basis for chemical and industrial manufacture meant that pharmaceutical companies could increase volume, multiply production, and minimize variation in their products.

These new developments, along with the formation of new multinational commercial entities, and the availability of new and high-value drugs, such as morphine, quinine carbolic acid, and salicylic acid, changed the entire nature of the craft of apothecary and gave birth to the modern scientific discipline of pharmacy. The advent of this new chapter in pharmacy required a shift in thinking about drug regulation. Public safety and public health had to be rethought in the light of these new developments.

Related to the advent of a new and rapidly growing pharmaceutical industry meant that the role of physicians in drug manufacture or drug design was becoming even more limited. While the physicians and apothecaries had been separated for some time, the influence of chemistry and new laboratory technologies created a bigger gulf between them. The new apothecaries were not just experienced herbalists, but also chemists who understood molecular relationships better than their physician counterparts. The clear separation between the physicians and the drug industry meant that physicians did not fully understand the role of each component within the drug and while they were the ones prescribing the medicines, their role in discovery, formulation, and compounding was now marginal.

The wide availability of the new medicines meant that the regulators now had more on their plate. The regulators had to think about more than just milling and grinding, baking and storing, but also about the harms of self-medication and advertisements. This was particularly true in the early part of the nineteenth century in the United Kingdom where arsenic was widely available as a rat poison, and was used frequently for seed dressing, treating sheep, and various industrial processes.[41] This led to the labeling of drugs and poisons through the Pharmacy Acts of 1852 and 1868. The act also mandated a separate storage within the pharmacists' shop. However, up until 1917, there was no restriction on the sale of poison by a pharmacist, the only requirement was that the buyer be known to the pharmacist, which itself was a subjective test.[1]

The nineteenth century also saw another major change in the kinds of therapeutics available to the physicians for prescription. A new class of treatment regimen, vaccines, were to arrive on the scene and forever change medicine, pharmacy, and public health.

After the landmark experiments of the British physician Edward Jenner in 1796, vaccines had slowly started to appear in the United States, United Kingdom, western Europe, and Russia as potent methods of treatment from deadly diseases.[42] Louis Pasteur's work added to the body of literature that further strengthened the case for vaccines. But it was not until the early part of the twentieth century that major pharmaceutical companies went full force into vaccine production and streamlining their processes. Vaccines had opened up treatment to new and previously fatal diseases. Microbiology got a real shot in the arm and became a major area of discovery and intellectual pursuit. Vaccines transcended the species, as they were being imported into the United States from Europe not only for human use, but also for use in farm animals. In 1870 the Boston physician Henry Austin Martin brought what he called true animal vaccine from Paris and brought "animal virus in tubes and on ivory points and squares of glass" that ushered a new area of animal farming in the United States.[43]

This rapidly changing landscape, with the arrival of vaccines and biological agents, meant that for regulators, new methods of testing had to be devised, which were different from chemical testing. This required not only development of new tools and strategies, but also training of a new workforce capable of doing comprehensive testing and sometimes exposing the regulators and inspectors to the high risk of catching the disease.

Another major development in the newly independent United States came in 1820, with the creation of the US pharmacopeia that was established by a group of eleven doctors.[36] Worried about charlatans and frauds, only 217 drugs were admitted to the first list of the US pharmacopeia. Concerned with imports and their unknown quality, one of the key goals of the US pharmacopeia was to ensure the quality of drugs being brought in to the United States through the empowerment of customs officials. This effort was largely unsuccessful because of lack of capacity among inspectors, to the extent that Jacob Bell, a British politician in the nineteenth century, remarked that "foods not good enough for the Brits were good enough for America."[44]

During the later part of the nineteenth century, with growth and innovation in agriculture, the bureau of chemistry in the Department of Agriculture under Harvey Wiley carried out one of the most detailed (up until that time) investigations of food adulteration and drug-quality testing.[45] Among his innovations was a bold move to create a dedicated workforce called the poison squad.[37] The poison squad was a group of brave volunteers, who acted as guinea pigs to test various foods by tasting them, much to their own peril, including long-term impairments. Wiley's efforts led to the passage of the Food and Drug Act of 1906, signed by President Theodore Roosevelt. Among the key provisions of the 1906 Act was the recognition of the US pharmacopeia (USP) standards and formulae as the official guide. It required that any drugs within the national formulary must adhere to the standards of the USP. In 1927 this bureau of chemistry became a separate agency called the Food, Drug and Insecticide Agency.[44]

Perhaps the most critical aspect of drug regulation in the twentieth century were public health disasters that led to the restructuring and empowerment of the regulatory agencies. In the previous chapter, we looked at the Sulfanilamide disaster in several states in the United States. It was truly a watershed moment in the US history of public health crises and drug regulation. Yet, ironically, it was not the fact that the drug itself had used an ingredient that was so devastatingly toxic. Instead, the real reason that the drug caused the deaths of so many people across the country was because of mislabeling. As the FDA historian Carol Ballentine points out, "Twenty-five seizures were made under the federal law. The charge was misbranding. 'Elixir,' FDA said, implied the product was an alcoholic solution whereas it was, in fact, a diethylene glycol solution and contained no alcohol. If the product had been called a 'solution' instead of an 'elixir,' no charge of violating the law could have been made. FDA would have had no legal authority to ensure the recovery of the drug and many more people probably would have died."[46]

This disaster and the realization that the FDA had limited legal authority led the commissioner of the organization, Walter Campbell, to say,

> It is unfortunate that under the terms of our present inadequate Federal law, the Food and Drug Administration is obliged to proceed against this product on a technical and trivial charge of misbranding. [The Elixir of Sulfanilamide incident] emphasizes how essential it is to public welfare that the distribution of highly potent drugs should be controlled by an adequate Federal Food and Drug law. . . . We should not lose sight of the fact that we had many deaths and cases of blindness resulting from the use of another new drug, dinitrophenol, which was recklessly placed upon the market some years ago. Deaths and blindness from this [drug] are continuing today. We also should remember the deaths resulting from damage to the liver that have occurred from cinchophen poisoning, a drug often recommended in such painful conditions as rheumatism. We also have unfortunate poisoning, acute and chronic, resulting from thyroid and radium preparations improperly administered to the public.
>
> These unfortunate occurrences may be expected to continue because new and relatively untried drug preparations are being manufactured almost daily at the whim of the individual manufacturer, and the damage to public health cannot accurately be estimated. The only remedy for such a situation is the enactment by Congress of an adequate and comprehensive national Food and Drugs Act which will require that all medicines placed upon the market shall be safe to use under the directions for use.[46]

The public outcry and the raw sentiment of anger, anxiety, and vulnerability, combined with the realization of the inadequacy of the legal code, led to the passage of the law that required more stringent and comprehensive safety tests before marketing.[47] However, it was not until 1943 that the federal Food, Drug, and Cosmetic Act was passed. This was followed by the Durham-Humphrey amendment of 1951. It was cosponsored by a former pharmacist turned senator, Hubert H. Humphrey Jr., who later served as the US vice president from 1965 to 1969. The other sponsor was Representative Carl Durham from North Carolina, who was also a pharmacist by training. The Durham-Humphrey act created two classes of drugs, namely over-the-counter and prescription drugs. The bill regulated prescription drugs and created stringent requirements to ensure that they could only be dispensed by the recommendation of a bona fide practitioner of medicine.[48]

While the elixir tragedy in the 1930s changed the drug-regulatory framework in the United States, it did little in the United Kingdom and the rest of Europe to create stricter enforcement. It was another tragedy, unfolding in the 1950s, that changed the European drug-regulatory agency's power and enforcement authorities. The European disaster was precipitated by Thalidomide, a drug used as a sleeping aid and for vomiting and morning sickness.[49] In the 1950s Thalidomide was doing brisk business in Europe. While the drug appeared on the market in 1956, it was not until 1959 in Germany and in early 1960s in the rest of Europe that deformations in extremities of newborns were noted. These deformations were characterized by long bones of the limbs, which in many cases had rudimentary function. Given the long bones resembling the seal's flipper, the disease was named phocomelia, coming from the ancient Greek word *phoke* for seal. The epidemiological connection establishing a direct link between the drug and the onset of phocomelia took several years. Up until that time, there was no law anywhere in Europe requiring the pharmaceutical companies to do premarket testing, including for drugs that may affect the development of human embryos. While several drug companies carried out the tests on their own, both on animals and humans, they were not required by law to do so. Additionally, these companies were not required to report or share those results with the public or with the regulators, even if they found something concerning in those tests. The Thalidomide disaster changed this situation significantly. In 1963, while discussing the situation in the parliament, Kenneth Robinson, a British Parliamentarian, noted,

> I come to my last main topic which is the control and safety of drugs. This is of course a subject which was thrust to the fore both in this House and in the public press a year or so ago as a result of the thalidomide tragedy. The House and the public suddenly woke up to the fact that any drug manufacturer could market any product, however inadequately tested, however dangerous, without having to satisfy any independent body as to its efficacy and safety and the public was almost uniquely unprotected in this respect.[17]

The Thalidomide tragedy formed the basis of the parliamentary committee in the United Kingdom on the safety of drugs. This committee created a framework for preclinical trial testing, premarket testing, and testing after the drug had been introduced in the market so that its activity, efficacy, and potential harm could be monitored and shared with the public. This committee led to the landmark Medicines Act of 1968 in the United Kingdom, which created a regulatory framework of testing with four pillars of safety, quality, efficacy, and supply. The Medicines Act came into force on September 1, 1971, and also brought further development of the British pharmacopeia and its publications.

Across the world, in both emerging economies and in developing countries, the desire to improve medicine quality has continued to become a more prominent political issue, yet differences in governance and culture, along with colonial history and legacy, have continued to cast a shadow on regulatory framework in these countries. In India, for example, the major regulatory framework comes from the colonial-era Drugs and Cosmetics Act of 1940, which regulated the import, manufacture, distribution, and sale of drugs in India.[50] To date, the major functions of the Central Drugs Standard Control Organization, the national-level regulator in India, and the State Drug Regulatory Authorities are governed essentially by the original 1940 act.[51] The 1940 act, which was further strengthened by the 1945 Drugs and Cosmetics Rules, was in essence a first-of-its-kind mechanism to regulate drug imports in India.[52] Given the nature of the pharmaceutical industry in India, which relied heavily on drug imports (particularly until 1970), the focus on drug exports historically has been relatively a minor component of the Indian drug-regulatory framework. The mission of the Indian Central Drugs Standard Control Organization includes specific strategies that "initiate in framing of rules, regulations and guidance documents to match the contemporary issues in compliance with the requirements of Drugs and Cosmetics Act 1940 and Rules 1945; to facilitate in uniform implementation of the provisions of the Drugs and Cosmetics Act 1940 and Rules 1945; and to function as Central license approving Authority under the provisions of Drugs and Cosmetics Act 1940 and Rule 1945."[53]

With the changing landscape in the pharmaceutical industry in India, over the years, new acts have been added, including the Drugs and Magic Remedies Act of 1954, which focused on regulating the advertisements of certain drugs as magic remedies and narcotic drugs. The Indian pharmaceutical industry underwent a major transformation and started to play a major role in creating drugs locally after the 1970 Indian Patent Act.[54] As a consequence, more recent debate in the Indian pharmaceutical sector has focused largely on the patents and intellectual property landscape and less so on drug-quality control. Chowdhury and colleagues, in a report, "Administrative Structure and Functions of Drugs Regulatory Authorities in India," argue that in the decades since the 1940 act "there has been a dramatic change in the sector, without a corresponding modification in the distribution of regulatory responsibilities. India has emerged as a manufacturing hub for generic medicine, and this requires greater regulatory focus and resources to be invested on manufacturing licenses and enforcement. However, both these functions are largely outside the purview of the central government and fall squarely within the competence of state governments."[50] They further add, that despite the additions and addenda to the 1940 act, there has been "an overwhelming concentration on subordinate legislation leading to an increasingly complicated system of rules that are difficult to tract and understand. This sector, therefore, is characterized by lack of legal certainty experienced by both regulates and regulators alike."[50]

The Indian angle on drug regulations, while still retaining many of its laws and legal constructs from its colonial past, is compounded by multiple factors. It is a combination of a highly populated country with substantial healthcare needs, a continued push to become the "pharmacy of the world," and increased demand from abroad for its low cost drugs. It is also a center of attention as a hub of generic pharmaceuticals, including the presence of international giants like Cipla. The battles for patents and intellectual property, and not just for drugs in India but also for drugs at lower costs for the poor in Africa, are often fought in the Indian courts.[55,56] Indian manufacturers, over the years, have continued to produce the bulk of their drugs, and some estimates suggest as much as 80 percent or more of drugs in India are manufactured locally.[57] Yet the enforcement of the standards of quality across the board have been applied with mixed success, and concerns about quality of drugs manufactured in India, for the Indian market as well as for exports, have continued to persist in recent years with a number of high-profile scandals associated with poor-quality medicines. Among the recent scandals was a guilty plea and a $500 million fine for the Indian generic pharmaceutical company Ranbaxy, which falsified data and sold subpar drugs, including medicines such as gabapentin, used routinely for epilepsy, to US consumers.[58]

Other countries, emerging out of the colonial era, have had a different trajectory. While the case for each country is different, the case for Ghana, the first black African nation to become independent, is an interesting example of postcolonial regulatory development. Although Ghana gained independence on March 6, 1957, it was not until 1992 that its Food and Drugs Act was passed.[59] Prior to the 1992 act, the framework was largely based on ad hoc measures, many colonial-era frameworks and structures, and on a loosely defined Sale of Goods Act of 1962. Act 137, or the Sale of Goods Act, focused on the physical safety of goods and distribution facilities and stated "that a contract of sale of goods is a contract whereby the seller agrees to transfer the property in goods to the buyer for a consideration called the price, consisting wholly or partly of money."[60] This regulatory framework, which captured general commerce practices, was largely inadequate for the needs of a country with a growing burden of infectious disease and issues associated with access, pricing, and quality of drugs. Some estimates suggest that 60 to 80 percent of the cost of healthcare in Ghana goes to drug procurement and providing access to medicines. In the light of these challenges the

Food and Drugs Act was created in 1992 and amended in 1994. The subsequent Pharmacy Act was passed in 1994.

The Food and Drugs Act of 1992, among other things, also focused on pharmaceuticals and pharmaceutical quality. In section 11, it "prohibited sale of drugs and other chemical substances, that may cause injury to the health of the user when used according to the directions on the label accompanying the article."[61] The Food and Drugs Act also created the food and drugs board (FDB) of Ghana, which is responsible for regulating pharmaceuticals. The responsibilities of the FDB included the control of manufacture, import, export, distribution, inspection, regulation, pharmacovigilance within the country, good manufacturing practices, collaboration with international agencies, and advertisement of drugs.

With the legal framework being very recent, combined with a lack of critical resources (human and financial), the burden of disease, and other challenges, has been a hallmark of broader challenges in a number of developing countries, not only in Africa. The Nigerian National Agency for Food and Drug Administration and Control (NAFDAC) was created in 1993 and while it has tried to create national structures and administrative zones, an increasingly aggressive campaign against counterfeiting, challenges with quality of medicines in the country have still persisted.[62]

Chinese manufacturing has taken the world by storm in the last couple of decades. This growth in manufacturing has also affected the global pharmaceutical market. China continues to play a major role not only in producing a large number of drugs but also in the active pharmaceutical ingredient, which is the key component of the drug, to countries like India, who themselves are major producers of the pharmaceutical products.

Broadly speaking, the drug regulatory framework in postcommunism China can be divided into two parts: pre- and posteconomic reforms.[63] The pre-economic reform period spans from 1949 to 1977. As part of the communist government approach to bring pharmaceutical industry, procurement, sale, and quality under a unified framework, a series of institutions were created. These included a state-owned Pharmaceutical Company of China, which was established in 1950 to deal with wholesale trade. Subsequently the Agency of Pharmaceutical Industry was created in 1952 to control the manufacture of pharmaceuticals and chemicals. To regulate and standardize practices in the widely used sector of traditional and herbal medicine, the Traditional Herbal Medicine Company was created in 1955. All of these state-owned enterprises were, in their own framework, also responsible for ensuring drug quality and were centrally controlled. In addition, another higher-level structure, called the Drug Supervising System, was developed, which belonged to the Ministry of Health. This supervising system was a framework to manage and maintain quality of healthcare commodities and services throughout the country. The various structures had similar goals and statutes that were often redundant.[63] Without a clear definition of responsibilities, these two systems, operating at the national and provincial levels, often failed to work in an independent and transparent manner. As the pharmaceutical companies in the pre-economic reform period were also largely state-owned, the regulators and those being regulated often had a complicated relationship.

The cold war, the Chinese foreign policy, and the nature of economic models in effect in China until 1977 meant that there was little interaction with the rest of the world in terms of the production of pharmaceutical products. Additionally, data were tightly controlled, if collected at all. It is therefore difficult to estimate the extent of problems associated with counterfeit or substandard medicines.

With reforms being introduced, the pharmaceutical sector also saw both rapid growth and new structures for its regulation.[64] The first effort in this regard was the creation of a pharmaceutical administration, or PA, as a government agency that was designed to promote the pharmaceutical industry in the new China. However, this agency failed to make any serious inroads due to new private companies and a desire by the state-owned companies to make all of their business decisions themselves, as opposed to working with the PA. The PA was never able to fully create a centralized regulatory framework, in part due to lack of government support, and in part due to its financial dependence on the government-owned pharmaceutical industry. While the pharmaceutical industry continued to grow at an exponential pace (for example, the number of applications for new drugs went from 10 in 1985 to 1,700 in 1995, and China had nearly 5,300 pharmaceutical factories by 1995), the increase came at a cost of declining drug quality.[63] A series of embarrassing incidents appearing in the international press, which questioned the integrity of Chinese-manufactured drugs sold in the United States and elsewhere, started to create serious internal pressure to create independent regulatory body within China to maintain drug quality.[65,66]

The first foundations of this independent body were laid in 1998, when the State Council of China created the State Drug Administration, or SDA.[67] The SDA was fundamentally different from the PA in many important respects. The SDA was not supposed to promote Chinese pharmaceutical industry, though the transparency of this institution has always been a subject of debate. The SDA, with time, morphed into a new structure, called the State Food and Drug Administration. This new agency, for a period of five years, underwent a new mandate and lost its independence from 2008 to 2013 and was required to report to the Ministry of Health. In 2013, this agency went through another phase of evolution, was rechristened the Chinese FDA (CFDA), and became an independent agency once again.[63]

As the Chinese products enter more and more markets around the globe, the CFDA has come under increasing pressure to improve the safety standards, ensure quality control, and prosecute those who have been complacent or negligent in their duties.[68] It has responded to this pressure by a number of high-profile punishments, including execution of a former head of the organization in 2007. There has also been a PR campaign to demonstrate, to international markets and stakeholders, that China is serious about both the quality of its products and public health. Yet a string of incidents in the recent past, including one right around the Beijing Olympics, which showed the presence of toxic contaminants in a number of food and pharmaceutical commodities, is likely to result in continued internal and external pressure on the CFDA.[69]

In the end, the regulatory framework for assuring quality in any given country is a product of three fundamental components. The first is the presence of incidents and crises that give rise to laws, regulations, and preventive measures. The incidents in the United States, Europe, and to a certain extent China have informed laws and have shaped the regulatory framework and the powers of associated administrative agencies. The second is a combination of the history and strength of the local pharmaceutical industry. In India and in Ghana, there is a strong presence of colonial-era regulatory frameworks that have been in place for decades and continue to still guide the legal framework and the power of the regulators. But in India, in contrast to Ghana, the presence of a strong generic industry has shaped the laws, which are not present in Ghana. The third component is international pressure, which continues to increase with an increasingly globalized world. Thus pressure on India and China along with recent engagements with the FDA are a sign of both the dependence of the United States on raw materials and finished products that are made in India and China, and an increasing global awareness.

Yet all of these factors rest on a simple underlying assumption, that the regulatory body and its staff are competent, able, aware, and of unquestionable loyalty. What if that very assumption does not turn out to be true?

5

Unaware, Unable, or Unwilling?

April 29, 2015. On a typical sunny day in Dakar, Senegal, I found myself in the busy streets of the bustling downtown. I was headed to Ker Sering Bi, a busy marketplace where, I was told, that you could find "anything," including all brands of drugs. Ker Sering Bi itself is a small place, tucked just inside the main road. I walked in. Old men in traditional garb were talking to each other while drinking tea. Their garments and hats reflected their traditional attire and the long and rich cultural history of Senegal. There was a sense of calm—the mosque nearby was calling the faithful to the noon prayer. It seemed like just another day in Dakar. I did not see anything suspicious. I walked further in, turning into the side streets. Here I found a long line of stalls, on carts, where one could find anything. From a mosquito repellent to CDs of the latest movies, from traditional toothbrushes (miswak) made from a mustard tree, to herbal remedies for skin whitening, everything was available and the business was booming. No price was fixed and everything was up for bargaining.

Something on one of the stalls caught my eye. As I paid close attention, there were brands of medicines that I recognized immediately, antimalarials and antibiotics of brands that I was familiar with, in dusty old boxes. My interest piqued. What were they doing here? I knew about pharmacies not requiring prescriptions, but this was more than that. This was free market in the freest sense of the word. I pulled out my cell phone, in the hopes of taking pictures for my research and to increase awareness. Just as I was about to take a picture, the stall owner jumped at me and started yelling. I did not recognize what he was saying but he was clearly very angry. It was about to get ugly.

I am neither an investigative journalist nor a professional photographer. I have no training or aptitude for any undercover operations or the ability to disguise my intent to get a big story. I also looked different from everyone else. There was no one in sight who looked remotely like me. My facial features and the color of my skin made me definitely not a local. Unlike Tanzania or South Africa, where there are plenty of people of South Asian background, Senegal did not have the same number of people from my part of the world. But I was also not quite the white-skinned foreigner whom locals were used to seeing. I did not even know any of the languages spoken here: the local dialect, Wolof, or the official language, French. I was handicapped by my appearance, abilities, and aptitudes. I did the worst possible thing—I tried to reason with the stall owner by telling him that I had come from America and that I was interested in knowing what he was selling. I am not sure why I did that: perhaps it was a rookie mistake; perhaps it was my sense of panic. In hindsight it was both extremely naïve and potentially very problematic. Fortunately, he did not understand a word of what I had just said. I was in no mood to get beaten up. So I gambled. I greeted him with a traditional Muslim greeting and asked how he was doing in broken Arabic. He did not understand Arabic but understood my greeting. Things started to defuse. I took advantage of this little respite and quickly walked away from his stall.

As I stepped away a young man in his mid-twenties approached me. He was wearing what may have been a Barcelona football club T-shirt at some point, barely recognizable now, and black trousers. He spoke little English—enough to get by, enough for me to understand what he was saying, and, more important, enough for him to understand what I was saying. I had traveled often enough to know that interacting with people like him could go either way. It can be very useful and insightful or it can get me into very serious trouble. Fortunately, it was midday, in the middle of the big street and with lots of people around. I engaged him. He told me that his name was Moosa—and he asked me what I wanted. He thought that I would like to get some "cheap" (cheap as in knock-offs) sunglasses, or bags "for the lady," or CDs. I told him, no, I was only interested in finding out more about medicines. In particular, I wanted to know where I could find cheap medicines of good brands. It took him a while to understand what exactly I was asking. We went back and forth a few times—but eventually he got it, or so he said. He told me to come with him and we went into an alley. We stopped in front of an old 1960s-era building that looked like a combination of an old house and a semirenovated office. There was no sign in front. He told me to wait outside in the alley, as he went inside. I waited outside, looking at my surroundings and making sure that I understood the exit route, in case I had to run. I saw trash on the ground, some open gutters in the distance that had a faint smell of sewage, and some roaming cats that were intrigued to see me. I waited for about five minutes. Moosa came back with another person who had a dark green plastic bag in his hand. He told me he had the drugs that I need. I was intrigued—I never told Moosa which drugs, in particular, I needed, but this other person seemed sure.

He carefully opened the bag, showing me boxes of drugs with explicit pictures on them that suggested that these pills were meant for male virility. He insisted that this is the best stuff out there, and he was willing to negotiate the price. I was not expecting this. I told him that is not what I wanted—I wanted drugs to cure malaria or bring down fever. He said I should take the drugs he had anyway, since they were the best stuff. When I showed no interest in that, he was disappointed. As we parted, he had words of advice for me. Not much business in what I wanted, he told me.

As I came out of the alley, Moosa took me to another courtyard. Men in lounge chairs were talking among themselves—probably back recently from the noon prayers. There was a clear hierarchy here. The oldest were sitting on the chairs. Their clothes were starched and seemed expensive. Those who were slightly younger than the elder group were sitting on plastic stools. The youth were standing and looking outside, as if surveying who was coming and going. The children were playing with the typical indifference to the world around them. It seemed like a family courtyard with several generations all engaged in their own worlds. Moosa told one of the youngsters what I was looking for. This young boy, in his late teens, nodded and then shared this information with someone slightly older—who told one of the seniors, who nodded again. The boy Moosa spoke to told us to wait, and then the boy went inside. We waited again. This time, I was not getting any strange looks—not sure why, but this seemed like somewhat of a more legitimate establishment. The elder men kept talking to each other.

We waited for what seemed like five or ten minutes. The young boy came back with a couple of dusty boxes. A number of branded drugs were in those boxes. They were not antimalarials, but some other drug with a name that I did not recognize. The active ingredient was also unfamiliar. I was not sure what the drug was for. Moosa knew some English but his technical knowledge was limited. We reached an impasse. The young boy told me that he did not have antimalarials right now, but he had other drugs that I could get. I was not interested in just getting samples for the sake of getting them. I was not even sure what was in these boxes. What if they had narcotics? I was not going to take any chances with the customs people, either in Senegal or back in the US. I did not buy anything from him, but he had a diverse portfolio of medicines that he was willing to sell me, without any questions or prescriptions, and at prices far lower than the pharmacy. Unlike my experience a half hour ago, this was purely a business transaction, done in full view, in a cordial manner. I thanked him for his time, took my leave, and moved up the street, back in the world of stalls, CDs, and t-shirts. I gave Moosa a tip for his translation services—he asked me if I needed anything else but he was also getting tired and had realized that I was not interested in the more lucrative commodities. I thanked him and he took off, disappearing into the same street he came from.

So how did I end up in Ker Sering Bi? How did I know where to find counterfeits and knock-offs? I had not come to Senegal to go on some covert mission. I was actually in Dakar to attend a conference organized by the International Federation of Pharmaceutical Manufacturers Association—an umbrella group of research-based pharmaceutical companies. Through common friends in the United States, I got to know the head of drug regulation in Senegal who connected me to one of her lieutenants. It was during my discussion with this employee of pharmaceutical regulatory authority that I found out about Ker Sering Bi. He told me that I could get anything there—anything that I desired, from antimalarials to antibiotics: I just had to go there. The fact that he, as a government employee, knew the exact location of the market for fake drugs was troubling on so many levels. Is the government complacent or complicit? Why don't they stop those who are dealing in this illicit trade? His answer, echoed by his boss (who was also in the meeting that I had in the morning), was that their hands are tied. Ker Sering Bi's business merchants are strong and powerful and have deep inroads into the government. Sometimes they are part of the government themselves. The government is not able to crack down on them. Another colleague in the meeting chimed in—defending the government—saying that the government has closed it multiple times, but each time the mushroom growth of illicit trade comes back when the government moves on to the next challenge. Fake drugs, I was reminded, was not the only problem the government of Senegal has to deal with. The merchants in Ker Sering Bi were connected to a strong religious order, had strong ties to both the government and to the grassroots, and wielded tremendous power. Government, I was told, did not have the means to exert pressure forever.

One thing was clear—the government was not unaware in Senegal. I had limited time at the conference and did not get a chance to go to Touba, a town about a hundred miles from Dakar and a spiritual center of the local Muslims and of a Sufi order called Mouridiyah.[1] Touba, it is said, is where the big trade is located. The ancient town of spiritual heads of religious order and pilgrims is flourishing and flushed with medicines—good and bad, but all unregulated. Touba has become the center of informal markets and has come a long way from a town of a few thousand in the 1960s to Senegal's second largest city.[2] The reason that Touba has become a center of trade and commerce is because of the combined effect of a special tax status provided by the government and the traditional inefficiencies of governance and regulation seen in many low-income countries. Also, because of strong spiritual connection and strong ties to the diaspora, who feel connected to Touba, the city has attracted, and continues to attract, serious external funding. This flow of cash and resources has also helped the city to become a center of commerce. The mixture of external funds, strong business culture, and poor regulation has all contributed to it attracting a variety of counterfeit commodities, including drugs.[3]

Touba was indeed the center, and unique in the size of its market of counterfeit drugs. But it was not the only place in Senegal that had this problem. Another argument for the widespread problem of poor-quality medicines, provided to me, was that the hospitals often did not have the medicines, the private pharmacies with quality products were too expensive, and finding cheap drugs in alternative markets was the only option for the poor. This is the same argument that has been echoed elsewhere, including in the *Wall Street Journal* story on the fake antimalarial medication (a drug called Coartem) in Angola.[4]

In my short meeting with the regulators and lab personnel, I got to see the layers of complexity of the problem from the eyes of the regulators and heard their stories. Their arguments ranged from corruption in the government to powerful business interests to the inability of the government to provide adequate medicines. But perhaps the biggest and most surprising argument was yet to come. I asked a provocative question to my colleagues in the drug-regulatory authority, about what exactly was in these drugs that are widely available in Senegal, and if they knew that they were actually fake? They said, in unison, that they had no clue what was inside them. I was not expecting this. I asked again—did the government or regulatory authority ever test some of the drugs that are being sold in Ker Sering Bi and other parts of the country? Were they potent? What was in them? I got the same answer—that they did not know for a fact. The reason was simple, I was told. The funds available to the government to test drugs come from international aid organizations from across the world. This aid, was almost always tied to a specific program or a specific ailment. The government of Senegal did not have the resources to broadly test the quality of all, or even most, of the drugs. So if the money from a particular aid agency or NGO comes for improving malaria outcomes, they can only test antimalarials—not drugs against TB or HIV or other antibiotics. They are not allowed to do so. Because the aid programs focus on a handful of diseases, but the drug market is large and diverse, they have no way of knowing what exactly is being sold on the market. Funds that are made available often come with strict guidelines as safety checks against abuse, mismanagement, or corruption. Little is available as discretionary funds. Just as sampling drugs and testing them outside the disease management and eradication program is hard, so is the use of funds for improving the quality of the labs, training personnel, and creating technical capacity to maintain the laboratory equipment. Functioning labs, with equipment, consumables, and personnel, would require a separate funding stream and plenty of bureaucracy. For example in Dakar, there is a national drug-testing lab. On paper, the lab had two high-performance liquid chromatography instruments, a gold standard to measure drug quality. In reality only one of them had been working and the other had not worked in a long time. Most of the funds that were available to the regulators were tied to a specific disease and focused largely on testing drugs that were sampled from government hospitals and clinics. So there was no way of knowing what was present in the drugs that fell outside the scope of this program. If the disease was not a priority, the drugs concerned with that disease were not going to be sampled or tested. It was as simple as that. Given these constraints in the funding, capacity, and scope of disease eradication programs, what I was hearing was not all that surprising anymore.

The second dimension in the role of government being complacent or complicit comes from a weak workforce. On a bright sunny morning in January 2015, in a leafy neighborhood with big houses and high walls, I entered the headquarters of the Kenyan drug-regulatory authority, known as the Kenya Pharmacy and Poisons Board. The neighborhood was dotted with tall trees and buildings that had checkposts and security guards. As I entered the building and informed the receptionist about our scheduled meeting, I was told that everyone I was scheduled to meet was still in a meeting and running late. I was told to wait until the senior officers became available. There were three sofas inside the building—all occupied by others who had come before us. I decided to sit outside and wait—the sun was strong, but not scorching. At around 11:15 A.M., the previous meeting concluded and I was ushered in.

As I went in, I was greeted by an unusual sight. There were boxes upon boxes, stacked mostly in neat rows, going from one end of the hallway to the other. The boxes had some numbers and dates on them, and their ceiling-high stacks had created a maze that was hard to navigate. The boxes created a new structure, almost like pillars, within the building and gave it a unique aura. As I moved through various offices and went to the conference room on the second floor, I was greeted by members of the inspection workforce, all of whom showed tremendous warmth and kindness.

I asked them the same questions I often ask in these meetings—does Kenya have a problem of substandard and counterfeit medicines? Where do the bad drugs come from? What is the government doing about it? How well equipped were the staff? What were their biggest challenges? The answers to many of those were standard, and identical to what we had heard elsewhere, including in neighboring Tanzania where I had been right before coming to Kenya. Yes, Kenya has a big problem, we were told—how big? No one knew. The drugs come mostly from outside Kenya via complicated supply chains. So what is your biggest challenge, I asked? In addition to corruption and vested interests from big and powerful businessmen, they pointed to the acute shortage of staff. The boxes that had created a maze within the headquarters were all samples waiting to be tested. Sometimes the test time is over a year—in a national lab that is understaffed and overwhelmed. The cost per test of a single pill can be upward of $100, I was told. The staff told us that if they found something suspicious, it was hard to prove in court—since the sample testing takes so long and a lot may have already been degraded. The government often did not release the funds for testing, and the lab therefore was unable to function. The private-sector testing had started to make up for this gap, but that testing was even more expensive and not quite capable of handling the supplies from all over the country.

Another challenge was the lack of staff on the ground, making sampling extremely difficult. The inspectors said that there are only a handful of them (fewer than a dozen) and they are all well known to the pharmacies. With a network of informers, the inspectors can be spotted from a distance, and the pharmacies close their shutters to avoid sampling. This may seem like a solvable problem, or at least a problem that should not become a bottleneck, but it is. With limited resources, high work burden, and no clear resolution of the problem, the morale was not particularly high. The inspectors, in general, are unable to sample adequately—and hence they are unable to fully capture the picture, and even if they do sample, it is often inadequate, or takes forever to get results. The only success they have had so far is when the drugs look suspicious—they change color or show visible signs of degradation, which represents only a small part of the big picture and is far from adequate.

The governance challenges in Kenya point to other problems that further complicate the picture. The first is the international connection, particularly with China. Kenya, like many other African countries, is greatly dependent on Chinese products for its drug supply.[5] Local manufacturing is relatively weak and makes up for a small percentage. The incentive to create manufacturing within Kenya is minimal and, with limited local capacity, is not easy to start.[6] Finding drugs that lack the necessary active ingredients or ones with subpar formulation would necessitate action according to the Kenyan law. But bringing Chinese companies to court is not only nontrivial it is also diplomatically challenging for a country that depends on Chinese aid, goods, and services for a vast number of projects. Without a clear testing protocol and long wait times for testing drugs in-country, the Kenyan authorities often do not have strong foundations from which to bring the case to Chinese manufacturers. The diplomatic angle is also something that the Kenyan government is conscious of, and it does not want to antagonize Chinese partners who contribute to the Kenyan economy and development in various ways. Further complication comes from the long and complex chain of actors who are involved in importing drugs, some of whom may be close to government, and bringing them to justice is anything but straightforward.

The registration process for Kenya, and many other African countries, is prone to major loopholes. In Kenya, at the time of registration, the manufacturer and the supplier are required to provide samples (of their own choice) for testing.[7] Upon certification that the samples do meet the criteria, the paperwork moves forward, eventually leading to a license for five years. However, because the government labs are overwhelmed, understaffed, and poorly functioning, sometimes the testing never takes place or is incomplete. The registration process often relies on the tests and analysis documentation provided by the manufacturer at the time of registration. Second, the samples provided by the manufacturer may be different, in quality, from what is eventually sold in the country. With little subsequent testing, it is impossible to tell whether the samples and the product match up in any real way. Finally, and perhaps most shockingly, the renewal of license does not require any further testing or providing of samples—it is done purely as paperwork and another five-year license is granted on a routine basis. As a consequence of these loopholes, there are companies that continue to operate in the country have either never been tested or the samples that they provided were tested decades ago. The only time something gets flagged in many countries is when an incident, such as a malaria drug shortage[8,9] or a tragedy involving the deaths of patients (such as the sterilization incident in India),[10] hits the news.

The problems in Kenya also reflect the regional challenges of gateway markets. Kenya is the gateway to East African markets, which means that drugs arrive in Kenya to be sold and consumed in countries other than Kenya.[11] These drugs, by law, would not require any testing in Kenya unless they were intended to be sold in Kenya. But the process gets complicated due to corruption and operators who break the law. Drugs marked for Uganda never make it over there, or if they do, they are smuggled back to Kenyan markets for higher profits.[12] These drugs are not registered in Kenya and escape any testing at the point of entry due to a combination of loopholes in the law and a lack of oversight by the government.

On either coast of Africa, in Senegal and in Kenya, the challenges are similar but also unique in many ways. The similarity exists in the extent of the problem and the very real risk to public health. The issues of lack of equipment, limited staff, and poor regulation are common between the two countries. Yet the Kenya challenge also provides a different dimension to the problem than what is going on in Senegal. Kenya being a gateway country has plenty of its own unique challenges. The presence of large unregulated markets was something that was present in Senegal but a smaller issue in Kenya. In both places, in Kenya and in Senegal, I asked a provocative question—"Is the government here complacent or complicit?" I never got a yes but never got a no either. It was clear that there was a certain level of frustration with the layers of bureaucracy, corruption, and in general a lackluster approach to addressing the problem.

The challenge of drug registration is not unique to Kenya alone. When I spoke to colleagues at the National Agency of Drug and Food Control (NA-DFC or BPOM) in Indonesia, a country which not only produces nearly all of the drugs for the local market, but also provides drugs for other countries,[13] a key challenge was the presence of nonregistered drugs on the market, which are outside the jurisdiction of FDA testing.[14,15] If a product is not registered in the country, and enters through illegal channels, the drug cannot be legally tested by the FDA and hence creates a separate set of challenges, needing investigation by other arms of the government that may be overwhelmed or may not have health as their top priority.

The question of the role of the government is not simply looking the other way, or not acting when there is mounting evidence of unethical activity or internal corruption. It is also about a complicated, opaque political system and legal framework that stifles real action. In many cases there is lack of a legal basis for action, even in more economically developed countries.[16] In many cases, the legal language is complicated and the law is unclear on what constitutes bad drugs. Then there is the issue of jurisdiction, including what can and cannot be done, and who has the authority to prosecute, as per the law.[17]

The situation in Pakistan is particularly enlightening in this regard. In Pakistan, the drug act of 1976 describes in detail what is meant by quality and creates a legal framework for making sure that the drugs in the country are safe.[18] Yet in practice it is an extremely inefficient system and few, if any, fully understand how the drug act is supposed to work.[19] Another peculiar development was the passage of the eighteenth amendment to the constitution, passed in 2010. This amendment is also called the political devolution act. Based on the amendment, the federal government is in charge of only a handful of ministries, chief among them are defense and foreign affairs. Everything else has been moved to the four provinces, including health and education. Numerous reasons have been provided for the devolution. They range from the original intent of the constitution written in 1973, which gave provinces maximum autonomy, to arguments about the dwindling cash reserves of the federal government that needed to pass the buck to the provinces. In reality, it has created confusion and new power struggles. Health policy and regulation has been affected largely in negative ways.[20,21]

As part of the implementation process of this amendment, the federal ministry of health was dissolved in 2010.[21] Health is now considered a provincial matter. Yet not all matters associated with health are considered provincial matters. After much confusion and debate, it was decided that regulation of drugs and medical commodities should be a federal matter. This resulted in the creation of the Drug Regulatory Authority of Pakistan, or DRAP.[22] DRAP was created in 2012 to manage issues related to drug registration, regulation, and quality. While DRAP was supposed to be a central authority, it soon became a center of power struggles within the federal and the provincial governments. Some provincial governments are still interested in creating their own version of a regulatory authority and are actively pursuing it.[23]

As part of the complicated DRAP act, the federal government and the provinces are supposed to work with each other, with information and data flowing between them. Yet the reality on the ground is far more complicated and confusing. On the ground, the provinces have their own inspectors and the federal government has its own—and there is no law or procedure that suggests or requires them to interact with each other, even when working in the same city or even in the same location. There is no clear incentive for the provinces to work with the federal government, share information, or come up with a coordinated strategy. Similarly, there is no central training manual or scope of work or standard operating procedures for the inspectors. The labs in various provinces are also not standardized for instrumentation, supplies, procurement practices, or even staff qualifications.[24] Even the central drug-testing lab, based in the southern port city of Karachi, is operating from an old food and grain storage building that is not a purpose-built lab to host high-end and sensitive equipment. In a country like Pakistan, where resources are scarce, a lack of coordination and mandate makes things even more inefficient.

The issue in Pakistan is not just lack of coordination in efforts to address the problem, it is also in lack of consistency in having the same message. Because there are no standard operating procedures, in the federal or the provincial system, there is no standard sampling method either. With a completely ad hoc sampling system, and poor statistical literacy among the staff, it can never be said with certainty what the quality of drugs in Pakistan is. The official government numbers are not based on any official published data. Some official numbers suggested that the percentage of spurious drugs in Pakistan was less than 0.5 percent of the total drugs sold in the country.[19] However, in 2010 the Minister of the Interior suggested that the number of fake drugs may be between 45 and 50 percent.[25] These numbers, even if they were to be believed, are likely to be based on information from public hospitals and do not reflect the drugs that are sold commercially in private pharmacies, which are a major source of drugs for the general public.

The problem is further complicated by another major loophole in the system associated with drug-testing practices. By law, products from external manufacturers, who operate outside Pakistan and provide the country with finished products, are not sampled and tested.[19] They provide samples at the time of registration, but during routine sampling (which happens fairly irregularly) these drugs are not tested. The law says that it is only the local manufacturers who are to be tested. Unfortunately even that happens at an ad hoc and infrequent rate, without any real schedule or a timetable.[22] With over half of drugs sold in Pakistan (in volume, and much higher in terms of the total percentage in terms of value) coming from outside Pakistan,[24] there is no testing mandated by Pakistani law of these foreign manufactured drugs once they enter the market.[18] This is problematic since the significant number of the medicines coming into Pakistan come from China,[26] which, in recent years, has had several challenges with substandard and counterfeit medicines. With strong economic and trade ties with China and new corridors of open trade becoming increasingly part of the Pakistan commodities portfolio, this major loophole is likely to make the problem worse. The China-Pakistan economic corridor (CPEC), worth upwards of US$50 billion, is considered a major achievement of the current government to open up new trade, industry, and development possibilities. With the newer investment initiatives between Pakistan and China through CPEC, the trade is expected to grow in all areas, including health commodities,[27] and the lack of the government ability (or in this case mandate) to test can potentially increase the number of unregistered, substandard, and poor-quality medicines.

There are other loopholes in the government framework as well. For example wholesalers who are a major part of the overall drug-supply chain in any country are never mentioned in the Pakistani law and are not part of the regular testing chain.[19] Thus, those who get supplies in bulk, such as active ingredients or semifinished products, are outside the framework of Pakistan's drug-testing act. This means that poor-quality raw materials can enter the system without any checks required by law.[28]

In a recent meeting at the DRAP headquarters, which I happened to be present at, the Pakistan Pharmaceutical Manufacturers Association (PPMA) was deeply upset at the DRAP leadership for increasing surveillance and tightening of drug-registration laws. Any effort to regulate, improve registration, and increase testing is being met with fierce resistance by the PPMA. The PPMA has recently taken a strong stance on the deregulation of DRAP[29] and has added to the continuing chaos in the drug-regulatory system. The government of the most populous and most prosperous province, Punjab, has also recently announced a provincewide campaign to improve drug quality, increase surveillance, and create new independent drug-testing labs.[30] Yet it remains to be seen how this effort would materialize, with the national and provincial elections just around the corner in 2018.

The registration problem seen in Kenya and Indonesia are also pervasive in Pakistan. Pakistan has the largest list of essential medicines (nearly four hundred) in South Asia. India is seven to eight times more populous than Pakistan, and it has a much larger pharmaceutical industry, yet the Pakistani list is much larger than the list for India or any other country in the region.[19,24] As a result of the extensive list of essential medicines, numerous drugs are listed with the same name and a number of drugs that are identical are listed with a different name. This leads to confusion and huge price differentials, making it difficult for the government to test various commodities and enforce its writ.

Finally a large fraction of the society relies on herbal, traditional Indian (ayurvedic), or homeopathic medicines. None of these commodities are part of the routine drug-testing protocol, and discussions are still under way on how to test and regulate those types of medicines. As there is a huge demand for these remedies, because of historical and cultural reasons, these drugs are prescribed widely, including by some medical doctors, who themselves are unaware of their chemical composition, side effects, or efficacy. Furthermore, many synthetic, biochemical, nonherbal medicines are sold under the guise of herbal, ayurvedic, and homeopathic remedies, thus bypassing the regulatory framework for testing. When a pharmacist is found to be selling a wide variety of drugs, sometimes for the same ailment, some of them can be tested, while others cannot, according to the current law.[19] Some drugs are local, and some are foreign; thus even for the same ailment, an inspector would be unable to sample all possible drugs, even if they are identical, simply because of their origin or whether they are labeled as synthetic drugs or herbal remedies.

Regulating herbal medicines, traditional therapies, and even homeopathic medicines is also not easy, because the manufacturing processes are varied and often not standardized and the producers of these remedies range in their form and function. Sometimes drugs are produced in large factories, sometimes in cottage industries, and sometimes mixed together on the spot in herbal clinics.

The case in Pakistan is not unique—similar challenges in poorly described legal framework, mandate, and limitations in testing ability make it nearly impossible to regulate medicines that are sold in several African countries with parallel systems in place.[31]

The problems of poor regulation, while more common in developing nations, are not exclusive to them either. Imports of poor-quality Heparin ingredients from China caused deaths in nearly a dozen countries, including more than eighty deaths in the United States.[32] More recently, a Massachusetts-based company cut corners, ignored regulatory guidelines, and sold tainted steroid injections that resulted in fungal meningitis. Sixty-four people in the United States died, and hundreds were hospitalized with fungal meningitis.[33]

The challenges of regulation, quality, and adherence to guidelines remain truly global.

It is not just individual countries and their regulatory frameworks, and inherent limitations in capacity, or intent, that is a cause for concern. Even at a global level, there is no agreement on what does or does not constitute a good drug. While it may seem trivial to differentiate a good drug from a bad one, there is little consensus. Historically, bad drugs have been divided into three broad categories.[34] The first category includes degraded (e.g., expired, or degraded due to storage, exposure to light, etc.) products. These degraded products are a consequence of poor storage and transport and also dependent on the integrity of the storage system, such as climate control and continuously available electricity. Vaccine monitors to detect exposure to any non-cold-chain environment are costly and difficult to implement in most cases.

The second category includes substandard products, which represent the broad group of drugs that are poorly manufactured, have less than the desired amount of active ingredient, are unable to release the active ingredient in the prescribed time (i.e., drug dissolution), or have other ingredients that may be injurious or deadly. This category is routinely seen in developing countries due to poor manufacturing and poor enforcement of good manufacturing practices, but it is rarely seen in higher-income countries due to stronger regulation, routine testing, strict penalties, and controls associated with importing raw materials.

The third and perhaps the most controversial category is for falsified or counterfeit drugs. The three terms (degraded, substandard and counterfeit) have often been used as synonyms but in legal and public health circles, they may mean very different things.[35] Broadly speaking, counterfeit drugs are category of drugs represents intentional falsification, meaning it is intended to deceive the consumer and the customer, in selling something branded as something it is not.

While these categories may seem obvious and straightforward, their nuances and details have raised controversy at the global level. Because "counterfeit" is a legal term, generic manufacturers sometimes feel that perfectly good products manufactured by them, with correct dosage, correct formulation, and correct dissolution would be considered counterfeit because of patent infringement. The legal battles, originating from the definition of counterfeit, and patent infringement, sometimes last for decades. The argument posed by generic manufacturers (largely from India, but also from Brazil) is based on the fact that they believe that drug intellectual property (IP) and patent laws are unfavorable for developing countries. These laws, these countries argue, stifle growth and the availability of cheap products—and it should only be the quality of the drug, its efficacy, and not its IP that should determine whether it is a public health hazard.[35] Furthermore, in India, the change in the patent laws in the 1970s meant that only the process, not the final product, could be patented. So if the Indian manufacturer could come up with a different process, but come up with the same product, as patented elsewhere, the practice was allowable by Indian law.[36] On the other hand, other countries, particularly the United States and some in Europe, have more encompassing patent laws that protected both the process and the product. This has led to continuous tensions among various countries with strong pharmaceutical industries.

This debate about what is a "bad" drug has led to a major global deadlock, with the World Health Organization (WHO) in its crosshairs. Two events in the late 2000s culminated in stifling the activity of an international drug-regulatory body, called the International Medical Products Anti-Counterfeiting Taskforce, or IMPACT. IMPACT was based on prior conversations in the WHO from the 1980s and 1990s, and the secretariat was created in 2006, with the main goal to create awareness, promote intersectoral coordination, develop technical competencies, and promote investigation and enforcement capabilities with regard to counterfeit medicines.[37]

IMPACT laid the foundations for becoming the global guardians of medicine quality and set out to define drug quality in clear, easy-to-use, and broadly agreeable terms. Defining quality meant describing what a counterfeit drug is. According to IMPACT "a medical product is counterfeit when there is a false representation in relation to its identity and/or source. Counterfeiting can apply to both branded and generic products and counterfeit products may include products with the correct components, or with the wrong components, without active ingredients, with incorrect amounts of active ingredients or with fake packaging. Substandard batches or quality defects or non-compliance with good manufacturing practices/good distribution practices (GMP/GDP) in legitimate medical products must not be confused with counterfeiting."[38]

While there were some murmurs about the legal mandate of IMPACT, nothing came to the surface for a few years. This changed drastically in 2008 and 2009 when Dutch and German customs authorities seized medicines manufactured in India, headed for Latin America, on the basis of patent infringement.[39,40]

This led to WHO member countries, led by India, questioning the legitimacy of IMPACT and WHO in enforcing patent laws and acting as a policing entity, as opposed to being a guardian of public health and safety. The argument presented by the countries objecting to the actions of the WHO and IMPACT was based on several factors. First, they argued that the WHO has no legal authority to seize shipments based on patent infringement. Second, since the WHO's mandate is to protect public health and well-being, there was no evidence that the drugs that were on the move were posing any harm to the public health of the citizens of the intended countries. They argued that "the fact that a product willfully breaches third parties' trademark rights is not a central consideration from a public health perspective. Enforcement and control therefore take on a different dimension and are unconnected to IP concerns."[41]

The conflicts among trade, IP, policing, public health, and quality that had been brewing for some time now surfaced at the biggest of the global public health stages.

The original definition of counterfeit drugs, while already objected to in the early 2000s by India, became the subject of intense attention after the incidents in Europe, with member-states seriously objecting to the legal authority of the WHO. They started questioning what the goal, scope, and mandate of the WHO really was. It created intense pressure on both IMPACT and the WHO to come clean and define its vision and its authority. The pressure led to the WHO director general, Margaret Chan, commenting that "the role of the WHO should be concentrating on public health, and not law enforcement."[42] This was a major step to both clarify the mandate and the scope of WHO activities as well as to avoid continued conflict between India, Brazil, China, and other generic manufacturing countries versus the European and US partners who wanted stricter patent rules applied.

As a result of this controversy, the WHO tried to reach a less controversial middle ground. The new accommodating strategy was to strike a balance with a new and more complicated definition called SSFFC (substandard/spurious/falsely labeled/falsified/counterfeit medical products) to designate the range of products that may raise public health concerns and require international cooperation for the purpose of their prevention and control. Despite this development of a newer middle ground, the member-states still disagree about the action plan and collaboration between various units of a new "member-state mechanism" approach that is focused to build cooperation.[41]

The increasingly complex global-supply chains, with products originating in countries far from the destined nations, combined with an unclear definition of good or bad drugs, and a lack of legal jurisdiction along the way make the problem both complex and frustrating for country regulatory bodies. As a result of this impasse, pharmaceutical companies and public health researchers are concerned about active solutions to the problem and there is increasing worry about whether a collective global solution, with both resources and commitments from a wide variety of international stakeholders, will ever be possible. There is a desire to create an international treaty among legal and economic circles, but little has been done in that regard to make that dream a reality.[43]

There is also a sense of frustration toward the WHO, which is unable to fully define its own role and mandate. Does the WHO safeguard the health and well-being of people, and does that imply that it should also try to regulate quality and protect products from counterfeiting? Or is that outside its jurisdiction and the trade laws, intergovernmental interactions, and legitimate trade in generic medicines should not be a concern for the WHO? The position of research-based pharmaceutical companies and their umbrella organization, IFPMA, is favoring stricter guidelines and a stricter definition; however, generic drug manufacturers are opposed to that. It is important to note that the generic manufacturers based in India and China to a great extent, and Brazil to relatively a smaller extent, do provide a large number of pharmaceuticals in Africa, have played an important role in the pricing of HIV drugs, and hence have considerable influence when it comes to issues of access and pricing.[44]

Yet somewhere in the middle of this debate, important questions are being ignored and extreme positions are putting both individuals and public health programs at risk. The issues of trade and IP cannot be fully dissociated from issues of quality. Roger Bate in his work analyzes this position for developing countries. He argues that "Brazil and India are correct to insist that anticounterfeiting measures not impede the trade in legitimate generic medicines, but by opposing anticounterfeiting measures too broadly, the same countries risk under-mining efforts to exclude life-destroying counterfeits from commerce, including where such drugs affect their own citizens or industries."[34]

Thus the challenge for a country like India is not just to protect its generic pharmaceutical industry, that makes up a significant part of its exports, but also to protect its own citizens who may be affected by a position that blocks any global effort against counterfeiting.

There are multiple layers of complexity in regulating quality, and each layer presents a unique challenge. The problems start from the level of individual nations that sometimes genuinely lack the human and technical resources and sometimes decide to look the other way. Countries sometimes even flatly deny that the problem exists within their borders, and they present data that are hard to believe. Reliant on international aid, some countries argue that their hands are tied to have a broad-based program. They argue that they can only focus on specific diseases and not broad sampling, testing, and strengthening of the whole system. This argument is only partly true, as there is also a lack of emphasis on engaging citizens, which may not require any external aid.

As if these problems were not hard enough, the problem when mixed with the legal aspects of jurisdiction makes the already difficult situation intractable. The issue of jurisdiction is not just a problem for individual countries. Countries at a global level continue to disagree on a very basic and fundamental question—what is a bad drug? And who is supposed to regulate quality?

How is one to solve a problem, if we cannot even agree on what the problem is?

That question remains unanswered.

6

Crime and Punishment

In 2014, as part of a small pilot project, my colleague Katie Clifford and I started collecting news stories of counterfeit, substandard, and poor-quality medicines from around the world. Using simple keywords, we created a small program that would scan the news and record anything appearing in the local or international press about the presence of substandard and counterfeit medicines. To make sure that we did not double-count and find redundant news at multiple sites, we had a few filters in place. Sometimes the news would be about a raid at an unauthorized facility, and sometimes the story would be about consumer complaints against a particular drug. Often, our search would find a story that was basically a chest-thumping exercise from the local government that wanted to highlight its commitment to public health and consumer safety. We found plenty of pats on the back.

Right from the very beginning, we knew that whatever we were going to capture from this small study would be a small fraction of the total data. We knew that our little program was far from comprehensive. This understanding was not based on some innate belief about the complexity and the size of the problem, but based on our own sampling methods. First, our searches were only in English and hence we were most likely undersampling what was being reported in other languages. Our stories were likely lacking news from West Africa, the Middle East, Latin America, Russia, and China. We also knew that not all newspapers have a presence on the Internet; what was not in the public domain or on the Internet was beyond our sampling. We also knew that stories from China, which could be embarrassing for the government, may be subject to strict censorship and often not reported. Yet it was a simple exercise to see how often something related to counterfeit or substandard drugs was reported in the news on the Internet.

This small project took on a life of its own. Despite significant undersampling, we noticed that there were nearly twenty unique stories every month. Sometimes a dozen or so stories would appear in just a week. Some regions had an occasional occurrence while other regions kept appearing over and over again in our searches. In a quarter, there were on average seventy to eighty stories from around the world. West Africa (particularly Nigeria), India, and China were among the top hits, but Western Europe and the United States were not entirely absent. Due to our language bias, we had very limited news from South America or Francophone West Africa, though there are strong reasons to believe that the problems are certainly present in those parts of the world. Similar arguments can be made about Portuguese-speaking countries in Africa and Latin America as well.

We then started parsing our data further to see any trends. In general, there were a lot more stories about counterfeits than about substandard drugs. On average, for every ten to twelve stories about counterfeit medicines, there would be only one or two about substandard medicines. This was largely because counterfeit medicines are easier to detect and make for a more exciting story than a substandard medicine would. Additionally, something that is outright fake and has dubious packaging and telltale signs of a fraud is easier to detect than something that has less than an exact amount of the active ingredient or fails to dissolve correctly.

In this small subset of information, the smallest component was made up of any news about criminal charges, the course of justice, or any penalties imposed on various parties. Any news that we would find about conviction would come from the newspapers and press in the United States, Canada, or Europe. Sometimes the crime would originate in a low-income country but the course of justice and punishment would occur in a high-income nation. This was recently the case for a Pakistani man from Karachi, Junaid Qadir, who sold medicines made in Pakistan to US consumers, claiming that the drugs were made in the United States. Qadir, who sold the drugs with under the alias Brad Pitt, was sentenced to two years by a federal court in Denver in 2016.[1]

A key challenge for my work since then has been to find out what happens when there is a clear case of fraud, misrepresentation, and malicious intent. How is that processed and prosecuted? How do the state prosecutors prove intent? What happens to those who are responsible? What is the legal framework for punishment and how severe is the punishment? The answers to these questions are complex, but possible to find in the United States, United Kingdom, and countries with a stronger regulatory framework. Even in the United States the penalties are relatively weak and usually run in the range of one to three years in prison, but only after strong evidence can be presented and intent to deceive or cheat can be proven.[2] The cases in many other countries, particularly in the developing world, where the problem is most acute and the societies most vulnerable, are far more complicated.

Armed with questions about intent, law, punishment, and deterrents, I asked colleagues in public health, law enforcement authorities, and departments of justice in countries where the challenges are most acute. In nearly all cases, and even despite strong evidence, the trails always ran cold.

It became obvious that our little experiment to document news was pointing to a problem in closing the loop. What starts off as a flurry of activity, and creates a juicy story, would often end up in the basket of no action. The lack of action against those who are involved in criminal activities also reflects the growing frustrations of so many of the stakeholders, not the least of whom are patients and caregivers.[3]

The problem is affected by cracks within the system, at every point, with cases of crime, fraud, and negligence. Even when a country or a region gets lots of bad medicines, only a small fraction of these can be detected because of challenges in technology, governance, and oversight.[4] An even smaller fraction reaches the courts, and by the time a fine or a punishment is meted out to the guilty party, it becomes a rarity.[5,6]

So why does this happen, even in scenarios when there is ample reason and evidence to support the widespread prevalence of bad medicines?

The answer depends on whom you ask. The previous chapters show that it is not an easy problem to track. Lack of technology, limited infrastructure, minimal training, and human resources that leave a lot to be desired make it very difficult to catch the perpetrators anyway.

Even when there is a clear case of fraud or negligence, proving the crime in the court of law is anything but easy, and it usually does not lead to conclusions that seem particularly punitive. In the case of Efroze, whose product led to the deaths of over two hundred people in Pakistan, there was a long process, with experts, international technicians, testing in London and Switzerland, and international tribunals being created to get to the heart of the matter.[7] The report by the World Health Organization (WHO) on this matter was clear in its conclusions of criminal negligence by the upper management, yet the case was bounced from one court to the next, reaching the supreme court of Pakistan.[8] Eventually, after long delays, the supreme court ruled that Efroze had to pay compensation to each person who died, as well as to those who were affected.[9] The amount agreed was 400,000 Pakistani Rupees (PKR) per person who had died as a result of the bad drug. This amount was less than US $4,000 per person.[8] Furthermore, the penalty was to be given out in twenty-four equal installments. This meant that the company only had to pay $150 per month for each patient for a couple of years. Quite simply, this is a drop in the ocean for a company that makes significant profits by selling its products all across the country. The cost for each patient who was affected was $20 per month to the company, and that too only for a total of two years. Perhaps a bigger story was that the company did not have to pay any penalties to the government or lose its license, and its senior executives did not face any serious disciplinary action. The company continues to all over Pakistan with no restrictions.

Despite the meager penalty imposed, the case of Efroze is still quite unique. First, it was a local manufacturer who had made the tainted drugs. If the drugs were imported from abroad, the case would have been much more difficult to prosecute. Second, in this particular case, there was a direct correlation between the product and patient fatalities. The presence of an antiparasite drug, pyrimethamine, in the drug, when it should not have been there, was neither disputed nor denied by the company staff or its executives.[7] Also, there was no dispute that the cause of mass casualties was a direct result of drug contamination. The fines imposed had little to do with the poor-quality manufacture, violations of code, or poor adherence to safety standards. This was about the deaths of over two hundred people. So the case was a lot clearer than most other scenarios, which are far more complicated. Had it been just the violation of regulations for proper manufacture and lack of hygiene and safety procedures, the fines would probably have never been imposed.

In most other cases, the seizure of bad medicines is in the market, in the port of entry, or as part of a sting operation in a factory that is making spurious drugs. Yet the next steps after seizure are fraught with complications. The challenge in what to do next is three-fold. First, a drug that is seized needs to be checked against the list of registered drugs. If it is indeed registered, it needs to move to the next stage of testing and beyond. If it is not registered, it falls in the gray zone of authority. In Indonesia, for example, this problem is particularly challenging. The country manufactures most of its drugs. A drug that is imported, and appears on the market without appropriate registration, would require a completely different legal channel than a drug that is on the market with license and is of poor quality. The jurisdiction battles are often long and confusing between various agencies of the government. With smuggling a problem, not just in Indonesia, but also in several other countries,[10,11] the list of drugs that are registered represent a far smaller number than the actual drugs on the market,[12,13] making the legal proceedings difficult to start.

The second challenge is about backlogs, poor resources, and training, which makes it difficult to bring a case to court with all the evidence. The storage conditions between the raid and the testing are often not of the highest standard, in terms of temperature control and security. The accused party can argue that the drug lost its potency or was tampered with due to poor storage while it was waiting to be tested in the national facility.

The third aspect is an international angle that makes the problem even more complicated. It is hard to prosecute an importer who is responsible only for in-country supply, when the drug itself was made without regard to proper regulations and practices in another country. The importer may not be required to do any testing by the laws in the country, as is the case in a number of developing countries. The case against the importing company is going to be inherently weak. Countries in Africa, Asia, and Latin America that rely largely on cheap generic drugs from the manufacturing powerhouses of India, Brazil, and China often do not have any political leverage in the manufacturing countries to bring the perpetrators to justice.[14] There are few treaties and no formal extradition agreements to pursue crimes across international borders. It can also be politically challenging with strong consequences for a country, in sub-Saharan Africa for example, that depends on foreign aid and investment from China, to bring this to international media attention and create diplomatic complications.

Perhaps the biggest challenge comes from the vague, and often absent, legal code that describes what to do in case a drug is found to be counterfeit or substandard. Often times there is no legal framework that is available to build a strong case. In Tanzania and Kenya, for example, the issues of counterfeit are dealt with anti-counterfeiting agencies that have more expertise in counterfeit electronics than in counterfeit drugs. In Pakistan, the Drug Act was written more than forty years ago and has vague guidelines on the punishment of counterfeiting or intentionally producing poor-quality medicines. While politicians often make public statements about using the full force of the law to appease their electorate, the exact nature of what can and cannot be prosecuted, what are the legal guidelines and how to enforce them, is anything but easy.

In my meetings with the Kenya poisons board, there was a general frustration about the existing legal framework. I was told that even if all goes well, and after laborious and expensive proceedings, the court does indeed find fraud or unlicensed operations, the fines are less than US$200. This was confirmed by recent newspaper stories appearing in Kenya.[15] A paltry fine of $200, after years of delay, would hardly be a disincentive to engage in illicit trade. What happens in the case of a repeat offense? What are the penalties for not keeping the licenses up to date? What are some consequences of intentionally selling expired products? The criminal code is often lacking in these scenarios. In Kenya, there is further confusion about how the devolution of healthcare services to the counties would affect the legal framework and the battle against substandard and counterfeit drugs.[16]

Even in the United States, there are serious challenges associated with prosecuting the crime associated with the sale of counterfeit medicines. The deputy commissioner for the Food and Drug Administration (FDA), Howard Sklamberg, in front of the US House of Representatives' subcommittee on Oversight and Investigations expressed his frustration in 2014 when he said that:

> the reality is that the criminal penalty for the risky and inherently dangerous practice of importing unapproved foreign drugs is simply not sufficient to deter the criminal element. The penalty for such conduct, which generally falls under the "misbranding" and "unapproved new drugs" provisions of the FD&C Act, is three years imprisonment, and only if the Government can show that there was a specific intent to defraud or mislead. Otherwise, it is a misdemeanor, punishable only by a maximum of one year imprisonment.
>
> The penalties for health and safety violations for distributing unapproved or misbranded drugs have not been revised in decades and are substantially less severe than penalties for violations relating to intellectual property or economic loss. Title 18 Counterfeiting, designed to protect the trademark holder, carries with it a 20-year maximum penalty for counterfeit pharmaceuticals. However, risky conduct such as

> trafficking in foreign unapproved or adulterated drugs, carrying with it the same risk to the public health, is subject to a one- or three-year penalty—same risk to public health, dramatically different results.[2]

Mr. Sklamberg provided the example of a Utah man who was convicted of "trafficking in internet sales of drugs" that were not approved for sale in the United States. He procured the drugs from international suppliers, and since they were never approved for sale in the United States and had already been distributed to consumers, it was difficult for the FDA to determine whether the drugs were counterfeit, substandard, or adulterated, or whether he was simply selling pharmaceuticals that had not been approved for sale. While the net sales by the Utah man were over US$5 million, the nature of the crime and the vague legal structure resulted in only a one-year sentence.[2]

It would be inaccurate to argue that there is no effort to address the legal shortcomings in prosecution. The Council of Europe, an organization committed to promoting law and human rights in Europe has tried to create an international framework to increase cooperation and prosecution of counterfeiters. Not to be confused with the EU, the Council of Europe in 2011 came up with a treaty that was called the Medicrime convention.[17] The goal of the treaty was to criminalize a variety of activities, including the "manufacturing of counterfeit medical products; the act of supplying, offering to supply and trafficking in counterfeit medical products; the falsification of documents associated with trafficking drugs and the unauthorised manufacturing or supplying of medicinal products and the placing on the market of medical devices which do not comply with conformity requirements."[18]

The treaty of the convention states as its core mission that it "provides a framework for national and international co-operation across the different sectors of the public administration, measures for coordination at national level, preventive measures for use by public and private sectors and protection of victims and witnesses. Furthermore, it foresees the establishment of a monitoring body to oversee the implementation of the Convention by the States Parties."[19]

Since its ratification in 2011 in Moscow, twenty-six countries largely from Europe as well as some from Asia (Japan, Israel, Azerbaijan, and Turkey), Africa (Guinea, Morocco, Burkina Faso), and the Americas (United States and Mexico) have signed on to the agreement. Despite its focus on crime and international cooperation against counterfeit drugs, there are major challenges associated with its impact on the ground. First, India, China, and Brazil, which are some of the major producers of generic drugs on the global market, did not sign the treaty. There is a greater emphasis on counterfeit drugs and falsified products and documents, which has been contentious between countries for quite some time.[20]

Second, the application of the law rests on countries actively devoting resources within their own borders and prosecuting those who are found guilty. That remains difficult to enact and enforce. The third challenge, which often gets ignored despite its tremendous impact, is how to pursue cases where the drugs are intentionally substandard and poor quality, or are a result of negligence or produced in environments that lack adherence to acceptable practices of manufacture.

Another international collaborative effort that has occurred recently is in the area of Internet commerce and its impact on pharmaceutical sales. The growth of Internet commerce has made an impact on not only online pharmacies but also the prevalence of fraud and proliferation of counterfeit medicines.[21] In response to these efforts, the international community has responded, with an effort led by Interpol and local national and regional law enforcement agencies.[22] The largest of these operations has been named Operation Pangea, named after the large supercontinent that occupied the planet during the late Paleozoic and early Mesozoic eras. Operation Pangea is an Interpol-led global effort that aims to tackle distribution of counterfeit medicines, illicit pharmacies, and online fraud.[23] Started in 2008 with initial participation from ten countries, the focus was combating fraud in Internet sales of drugs. Partners in the first Operation Pangea included the Permanent Forum on International Pharmaceutical Crime, Interpol, and the WHO's International Medical Products Anti-Counterfeiting Taskforce (IMPACT) initiative. The idea was to coordinate an "Internet Day of Action" where multiple partners would act together on a single day at multiple locations.

The main goal of the first operation was to tackle those "individuals behind Internet sites which illegally sell and supply unlicensed or prescription-only medicines claiming to treat a range of ailments." On the particular day, "locations in each country were identified, with investigators visiting residential and commercial addresses relating to Internet sites believed to be selling unlicensed or prescription-only medicines claiming to treat conditions such as diabetes, obesity and hair loss."[24]

The success of the first operation prompted creation of a second one. Given the size of the challenge, Operation Pangea II in 2009 moved from a day of action to a week of action, and the number of participating countries increased from ten to twenty-four along with partners in the regulatory agencies that included the UK Medicines and Health Care Products Regulatory Agency (MHRA), the US FDA, Immigration and Customs Enforcement (ICE), as well as the Royal Canadian Mounted Police and Health Canada. The operation, compared to 2008, was significantly larger and resulted in the identification of 751 unique websites engaged in illegal activity, out of which 72 were taken down. Interpol also announced that "in addition, more than 16,000 packages were inspected by regulators and customs, 995 packages were seized and nearly 167,000 illicit and counterfeit pills—including antibiotics, steroids and slimming pills, confiscated. A total of 22 individuals are currently under investigation for a range of offences including illegally selling and supplying unlicensed or prescription-only medicines."[25]

The operation has continued every year since then and the most recent one in 2016, Pangea IX, was conducted in summer 2016. The number of countries participating swelled to 115 from around the world. Additional partners from the private sector, including PayPal, Visa, and MasterCard, also collaborated in the latest operation. Pangea IX resulted in the launch of 429 independent investigations. As a result of this operation that was coordinated with social media, nearly 5,000 websites involved in trafficking illegal pharmaceutical products were taken down. Through the operation and coordination with customs, nearly 334,000 packages were inspected by various participating regulatory agencies and nearly 170,340 of these packages were seized. The operation also resulted in the seizure of nearly 12 million fake and illicit medicines and 393 arrests. The overall cost of the medicines that were seized from around the world were over of US$53 million.[26]

Despite the grand scale of Pangea, coordination between various governments, and a partnership between the public and the private sector, there are some criticisms of the operation. The major criticism stems from the fact that most of the seizures were in the United States, Canada, and Western Europe, places where the regulatory system is already strong and the impact on public health due to counterfeit and substandard drugs is smaller than in poorer parts of the planet. The second criticism is the fact that Pangea's main focus is Internet commerce, a significant problem in high-income countries but a smaller problem in rural settings in low- and middle-income countries that may be the most vulnerable. In these places most people typically do not buy their commodities online. The third criticism comes from the multifaceted challenge of substandard drugs that would go under the radar of Pangea with its focus on illicit Internet trade.[14]

Operation Pangea is not the only effort from Interpol to tackle illicit pharmaceuticals. There are regional programs that are tailored to the regional challenges. In Africa, Interpol has carried out operations named Mamba, Cobra, Porcupine, and Giboia.[22,27]

Operation Mamba started in 2008 in Tanzania and Uganda, and subsequently expanded to Kenya, Burundi, Rwanda, and Tanzania (mainland and Zanzibar). The latest Operation Mamba was Mamba III in July and August of 2010 and resulted in the seizure of 200,000 pills, 78 court cases, and 34 convictions. Given the scale of the problem in East Africa, this operation is relatively small.

Operation Cobra focused on seven Western African nations, including Burkina Faso, Cameroon, Ghana, Guinea, Nigeria, Senegal, and Togo. The last operation was in 2011.[28] While 110 individuals were arrested, including street sellers, there is little subsequent information on court cases or convictions. Poor-quality medicines and those who engaged in manufacture and distribution in these countries were targeted again in the larger Operation Porcupine that occurred in 2014 and deployed 2,200 officers. This operation resulted in the seizure of 196 tons of illicit medicines and led to over a hundred arrests.[29] A more recent operation, Operation Heera, focused on the region, particularly in Benin, Burkina Faso, Côte d'Ivoire, Mali, Niger, Nigeria and Togo and used nearly 1500 law enforcement officials to seize 420 tonnes of illicit pharmaceutical products.[30]

Operation Giboia is more recent and focused on seven Southern African nations, namely Angola, Malawi, South Africa, Swaziland, Tanzania, Zambia, and Zimbabwe. The most recent incarnation of this operation was conducted in 2015 and resulted in the arrest of 550 individuals and a shutdown of 20 pharmacies and 3 illicit factories. Approximately 3.5 million dollars' worth of medicines were seized during Giboia.[31]

To address challenges in Asia, Interpol launched Operation Storm, which focused on countries in South and Southeast Asia, including Afghanistan, Cambodia, China, India, Indonesia, Laos, Malaysia, Myanmar, Pakistan, Philippines, Singapore, Thailand, and Vietnam. Compared to African operations, Storm has been carried out fairly regularly with the first effort launched in 2008 and the most recent, Storm VI, in 2015. Here the focus was on a combination of online pharmacies and physical markets and pharmacies.[32]

The sizes of these Interpol operations vary, and so does the frequency of action. The last operation in East Africa, for example, was more than five years ago. While the effort by Interpol is an important step in multilateral collaboration, the overall scale is small in relation to the enormity of the problem. The efforts by local authorities, from law enforcement to courts, in aggressively pursuing the cases are needed to address the problem comprehensively. The evidence of focused, concerted, and sustained efforts in that regard is hard to find.

Outside the Medicrime convention and the Interpol efforts, there are other efforts in countries to create harsher penalties. Both India and China, as major global players of manufacture, and under international pressure to improve their quality, have rushed to create stiffer penalties for counterfeiting. Both have proposed using death penalty as a punishment.[33,34] In India, this was subsequently changed to life imprisonment,[35] whereas in China, death sentences have indeed been carried out. These include execution of six Chinese traders who exported fake antimalarial drugs to Nigeria in 2009.[36] A more prominent case of capital punishment was that of Zheng Xiaoyu, who was a former top food and drug regulator and found guilty of taking bribes to approve untested medicines. He was executed in 2007.[37]

Despite the harsh penalties, the problems of substandard and counterfeit drugs manufactured in India, China, and elsewhere have continued. The complicated trail of the manufacture and transport, the vested interests and corruption at multiple points, the lack of evidence to prove criminal intent, and the vagueness of the law that is broken by individuals and companies all make the tale of crime and punishment a murky one.

7

Trust and Mistrust

In 2014, after a long day of seminars and meetings at a conference where several pharmaceutical executives were present, I reached out to a senior Merck executive. I introduced myself and told him about work I was doing, both in the context of public health and in developing new technologies for point-of-care testing. I told him about growing up in Pakistan and why I believe that the technology gap is glaring and particularly problematic. My argument was that what was needed were solutions that were both contextual and affordable. He looked me straight in the eyes and asked me if I knew what was his biggest anti-counterfeiting tool? I was intrigued. “Tell me,” I said. “The Merck brand,” he responded calmly with immense confidence.

In business, trust in the brand is central to the core mission. How one builds that trust can vary, but as with my own experiences growing up in Islamabad, I knew that trust, while not always measured in raw dollars, continues to be a central pillar of a successful business. The moment someone starts to believe that your product causes harm, and is no longer a trusted commodity, the business starts to take a nosedive. Any sampling of television commercials of commodities from toothpaste to pain medicine would tell us that the advertisers often use the word “trust” in their commercial. It may come in a statement like “patient’s most trusted source” or “a doctor’s most trusted brand” to build confidence.[1] Trust is also the reason that efforts by large pharmaceutical companies against counterfeiting often fall in what they call "brand protection"[2]—a term that implies that the company wants to make sure that the brand name is not associated with commodities that are anything but trustworthy.

The relationship of pharmaceutical companies with global anticounterfeiting efforts often raises lots of questions. There are obvious economic reasons for big pharmaceutical companies to care about the integrity of their products, and they certainly do care about those aspects of the supply chain. But there are other questions that linger as well. For example, as global players that continue to highlight their commitment to global well-being through research and innovation, should we expect more from the pharmaceutical companies? Should they only care about the integrity of their own supply chain, and the moment a shipment or a drug leaves their purview, is their job done? And if there is indeed more that we can expect of them, what is that "more"?

Before any analysis of the position of the pharmaceutical industry, both large and small, generic or a research-based multinational, it is important to also understand the common perceptions that exist in low- and middle-income countries, where the reputation of large research-based pharmaceutical companies is complex.

The perception question is an important one, because it is related to trust in the commodities produced by the pharmaceutical company. By and large, the pharmaceutical industry does not enjoy a healthy reputation[3]—from the movie and the novel *The Constant Gardener*[4] to patent wars and price hikes—there is a sense that at the core of the pharmaceutical company, particularly a foreign pharmaceutical company, is a fundamental drive for profit.[5] The argument presented by those who favor this line, is that large research-based pharmaceutical companies are not interested in improving the human condition, they are driven solely by the profits. This position has been further highlighted in the prominent patent battles of the recent past, including the Novartis Gleevec case in India that lasted nearly a decade and went all the way to the Supreme Court of India.[6] The Indian Supreme Court decided against Novartis and upheld the decision by the Indian Patent Office that had rejected the patent application of Novartis. The patent issue has also been taken up by other institutions, including prominent international organizations, such as Médicins Sans Frontières (MSF, also known as Doctors without Borders), that have argued against extensive patent holdings by research-based pharmaceutical companies that affect generic manufacture.[7] The court battles, public opinion in the press, and the presence of large international NGOs all on one side have done little to improve the general perception of large research-based pharmaceutical companies.[8,9]

While the perception issue is certainly present and palpable in low- and middle-income countries, recent events in the United States associated with price hikes of the EpiPen[10] and a drug for HIV patients, Daraprim,[11] which increased its price from $13.50 to $750 per pill, have also affected the general perception about the motives of pharmaceutical companies in higher-income countries. These recent episodes have led to a broader discussion about ethics, the core interests of pharmaceutical companies, and the role they play in society.[12]

Another belief in low-income countries is that foreign, multinational pharmaceutical companies are complicit, or at the very least complacent, in the substandard and poor-quality medicines problem. The argument is based on a combination of mistrust, colonial heritage, and conspiracy theory. The proponents for this argument propose that because of the regulation gap between the rich and the poor countries, and the enforcement challenges, the pharmaceutical companies would often send their bad batches, or drugs that are about to expire, to poor countries, because they cannot sell them in the so-called developed nations.[13] There is also an argument for enrolling poor communities in high-risk clinical trials.[14] This perception, while seemingly far-fetched, is backed by some data from a few instances. In particular, there have been reports of drug donations in conflict zones or for humanitarian support that have been found to be close to expiry date or have expired already.[15] The World Health Organization (WHO) has had to issue press releases on this and has continued to lament the fact that this practice, while unethical, continues unabated and beyond the control of WHO.[16] Apart from a few incidents, there is, however, little evidence to prove that this practice is normal, widespread, or routine.

Proponents of the complacency argument would also point to the fact that because of weak regulatory structures, poor countries continue to import expensive drugs, under pressure from powerful pharmaceutical companies.[17] However, there is little, if any, evidence for a systematic supply of poor-quality medicines from large multinational companies to poor countries with a clear intention to deceive the local population.

The second complicating factor in this argument is the heavy baggage of colonialism. The long shadow of colonialism clouds the judgment of citizens who view the western corporate institutions with suspicion. Because there is little, if any, active monitoring of the drug-supply chains, particularly at the customs and arrival points, any conspiracy theory that is being floated about bad intent is often hard to refute due to lack of robust local evidence. That said, the colonial history and its impact are both very real and a very sensitive topic.[18] The exploitation for hundreds of years, through big businesses, guised in various developmental projects and efforts to "civilize the natives," has created a certain environment of mistrust that becomes a breeding ground for new conspiracies, regardless of any evidence to support it. For some in the developing countries, the pharmaceutical companies are an extension of the colonial era, ready to make profits at the expense of the poor, defenseless people. The neocolonialists, according to them, are not waging a traditional war, but a war of economic exploitation.[19]

Another dimension of complexity comes from an inherent distrust in the local government,[20] as well as with bureaucrats dealing with issues of health, customs, regulation, and quality control of commodities. Once again, while the clear link between large pharmaceutical companies trying to mislead the general public through bribery of the government officials is hard to find, the continued cases of corruption in the government sector make this link, even without evidence, easy to construct and proliferate among the general public. With the lack of social development in general, and the lack of accountability in the government sector in particular, the trust in government institutions remains largely low.[21] The government, bureaucrats, and the western pharmaceutical institutions are therefore viewed as colluding partners that add to the toxicity of the discourse.

The counternarrative, focusing on issues, evidence, and a clear position of pharmaceutical companies, is simply not discussed adequately or presented in an objective light. The pharmaceutical industry has also serious ways to go, to improve their perception. They have not been able to fully demonstrate that they care about health and well-being and are fundamentally interested in improving the quality of medicines, not just selling their own products. While there are some efforts on social media and creation of targeted programs for increased access, there are still major gaps in public engagement and improvement of perception. Pharmaceutical companies have not adequately addressed or countered the notion of "outsiders" interested only in profits and not in health in low-income countries.

The narrative battles for large, multinational, research-based pharmaceutical companies also take a hit from the challenges faced by them through generic manufacturers. The local competition, particularly in India, China, and other emerging economies, combined with long and public court battles on patents (that are not fully understood by the general public), further add to the trust deficit. The court cases on patents, such as recent ones in India, often move the conversation from specifics of the case, the details of the law, to the broader theme of exploitation and further complicate the efforts to address the issue in a comprehensive way.

In India, the rise of generic pharmaceutical companies with lower costs, local ownership, and a strong presence in the public sphere, as well as an increasing global presence in the HIV sector in Africa, have created a new sense of national pride in generic manufacturers.[22] Additionally, statements such as that by the former CEO of GlaxoSmithKline to say that generic manufacturers were pirates and that they have never done a day of research in their lives[23] did not go well. The response from the chairman of the Indian pharmaceutical giant Cipla, Yousef Hamied, summed up the sentiment of the Indian generics when he said that "if we're pirates, [let them] litigate against us. Where is the question of piracy when we abide by the laws of the land? I believe in patents. But I don't believe in companies holding a monopoly."[24]

In the backdrop of these challenging perceptions among the general public, there are several questions being asked of large multinational and research-based pharmaceutical companies. The first and perhaps the most important one is about the intention beyond the bottom line.[5] Is the large, multinational, research-based pharmaceutical industry interested only in brand protection? Or do they also care about the bigger questions that are associated with system strengthening, substandard drugs, and products that degrade within the system?

I have had similar questions and I have put them to the executives and researchers at Merck, Pfizer, GlaxoSmithKline, and most importantly the IFPMA. IFPMA stands for the International Federation of Pharmaceutical Manufacturers Association, a group that represents research-based pharmaceutical companies and associations around the globe in their efforts to build bridges across the industry, governments, and more recently with the general public.[25] IFPMA came into being in 1968 and over the last five decades, through policy documents, partnerships, and advocacy campaigns with global organizations such as the United Nations, WHO, etc., IFPMA aims to improve global health through better regulatory frameworks, innovation, and a stronger dialogue. The reports from IFPMA point to the commitment to the issues, including building patient trust in the vaccine and pharmaceutical industry.[26]

IFPMA also issues code of practice, what they term as "a model of self-regulation of pharmaceutical industry's activities in medicines promotion, communication and interaction with key stakeholders such as healthcare professionals, medical institutions and patient organization. Although self-regulatory, the IFPMA code is not voluntary and it is a condition of membership to the IFPMA for both member companies and national associations."[27] The first code was established in 1981 and over the last three decades it continues to be revised. It allows IFPMA to create both a structure for engagement and a guideline for its members on what they consider allowed and disallowed practices.

The arguments presented to me, both by individual research-based pharmaceutical companies and by senior leaders at organizations like IFPMA, on why big pharmaceutical companies care about counterfeit drugs, rest on several pillars. The first is the obvious argument about the protection of their brand and their reputation, which are important and essential pillars for these publically traded companies. This is, once again, about trust, which in their case translates into revenue, market share, and profits. Pfizer is not only worried about counterfeit Viagra, which seems to be widely available over the Internet, cutting its revenue stream; it is also worried about the adverse reactions it may cause to the patients. A negative perception would mean that people would lose trust in the real product or, worse, develop a negative impression of the whole company. Of all the counterfeiting that happens against Pfizer products, nearly 80 percent is Viagra related.[28] With the wide appeal of the drug, and the ability to buy the drug anonymously over the Internet, this poses a very real challenge to Pfizer's brand and its ability to maintain the trust of its customers.

But on a different level, the argument is connected to the issue of fairness and the rule of law. The pharmaceutical companies, like any other industry or individual, do not want to see someone else without proper authority produce their products, mislabel their commodities, or mislead consumers. They also point to the fact that their insistence on protection of intellectual property is in fact the defense of innovation, research, and scholarship.[29] If there is no intellectual property protection, there will be little incentive to innovate, they argue. They are worried that there is a risk to intellectual property as a whole in the developing world.

As a result of this consideration, there has been a major push from the industry to better define trade agreements and to enforce the stipulations of the World Trade Organization's Trade Related Aspects of Intellectual Property Rights (TRIPS) agreement.[30,31] The multinational research-based pharmaceutical companies want both to protect intellectual property but also work against trademark infringement that they are worried is under threat with more and more sophisticated counterfeiters. As a result, the technologies being used by pharmaceutical companies tend to focus largely on detecting frauds.[32]

Another argument from the research-based pharmaceutical companies is based on an unfair competition through generics, which continue to infringe on patents and make profits without having to invest in steep R&D costs, and without exposure to the inherent risk of failure that comes from investing in fundamental research.[33] Among individual pharmaceutical companies and their organizations, there is growing frustration with local governments and the local laws. There is also a strong sense that counterfeiters in a number of developing countries are not pursued aggressively, and weak policies, and even weaker enforcement of policies, are affecting genuine trade and commerce in quality products, including pharmaceuticals.[34]

The exact estimate of loss due to patent infringement and counterfeiting is unknown, but it is believed to be in billions of dollars and hence of major concern to the pharmaceutical industry. Beyond the labeling, licensing, and branding issues, there is a public health concern as well that is often cited by executives and

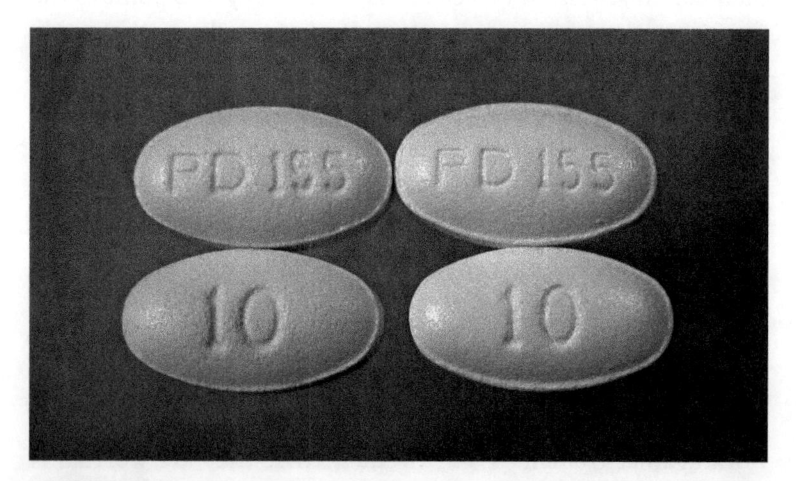

FIGURE 7.1.

Large pharmaceutical companies are worried that fake medicines not only undercut their business, but also erode consumer trust in their authentic products. Lipitor, a drug used to control cholesterol, is one of the most successful products from Pfizer. Here a counterfeit pill (left) is shown next to an authentic one. Reproduced with permission from Pfizer.

IFPMA through its publications,[26] and more recently through its online and social media campaign.

Research-based pharmaceutical companies argue that if a product is produced by a company that does not have the patent, does not spend seriously on R&D, and operates in a country with poor testing and regulatory laws, there is a strong chance that the final product may be hazardous and negatively impact public health. This is particularly true for new classes of drugs that are difficult to make and require extensive experience and technical expertise.[35] Pfizer is particularly worried about the widespread presence of counterfeit Viagra that may contain rat poison or may be coming from places that have not followed any hygiene regulation whatsoever.[36] Even when these products have some active ingredient, and do not cause any major incident, the issues in drug resistance, gradual poisoning, and subtle negative effects cannot be underestimated.

The pharmaceutical executives I spoke to stated clearly that they wanted to operate in environments where bad behavior is punished and that there were strong checks and balances against breaking the law and incentives to follow the rule of law. The executives argue that by not applying harsher penalties, those who spend a lot of money making good products are at a disadvantage against those who cut corners in design, manufacture, storage, and supply.

Multinational research-based pharmaceutical companies argue that the quality of their product is the same regardless of the country that they serve in (though prices and access may vary, as it depends on a lot of other factors and can create some discontent). In other words, the antimalarial drug Coartem, one that is genuinely manufactured by Novartis, will have the same quality in Zimbabwe as it would in Ghana or anywhere in Europe. To maintain quality in all of their products, these companies heavily invest in manufacturing, quality control, and quality assurance. While no company can argue that every single product, every single time, will be perfect, they say that by and large their products meet the strictest quality standards anywhere in the world. They also argue that bad practices, poor-quality medicines, and substandard drugs rarely link to systematic practices within research-based pharmaceutical companies, which are built around quality assurance and put trust in their brand as a central tenet of their business practices.

The same cannot be said about generic manufacturers, according to research-based pharmaceutical firms. They argue that with few checks and balances to ensure the integrity of the process within generic firms,[35] and in the absence of quality control at the government or regulatory level, there is an incentive to have poor-quality products manufactured at poorly run generic companies.[37] Without checks and balances, these products, made available in the market, are bad for legitimate business and much worse for public health.

While there is little doubt in the validity of these arguments, public health experts sometimes ask a deeper question about social responsibility. Are large research-based multinational pharmaceutical companies interested in resolving the problem of poor-quality medicines only for their own gains, and focus largely on brand protection, or are they genuinely interested in public health and well-being of all people. In other words, is the interest limited to making sure that only those who have the license and the authority to sell and make products operate, or is it genuinely centered in seeing the system improve as a whole?[5]

I put this question to lawyers, managers, and even research scientists in various multinational research-based pharmaceutical companies. There was general agreement that research-based companies do care, and they do so in ways that go beyond immediate profits and focus on long-term human development.

The research-based companies would point to their global commitment and significant investment in global health programs such as Merck for Mothers, a ten-year $500 million initiative to improve maternal health.[38] Other companies, such as Glaxo Smith Kline also point to their initiatives with their investment in global innovation.[39] New programs are created by Novartis to increase access and affordability.[40] IFPMA would point to their efforts in strengthening regulatory frameworks throughout the developing countries and would point to the series of meetings, symposia, seminars, and training they have offered for the regulators in low-income countries.[41]

Interestingly, many large pharmaceutical companies see themselves as the victims of malicious media campaigns, conspiracy theories, and assault on their creativity and the fruits of their hard labor.[42] There is a sense of frustration with local governments as well as laws and practices that are unfair in places where generic manufacturing is doing a booming business. This frustration is in sharp contrast to the perception on the ground, where research-based pharmaceutical companies are anything but the victim.

This gulf of perception is wide and brimming with mistrust.

The generic companies view the problem differently. The large generic pharmaceutical companies make several arguments in the defense of their practices. One is an argument about the high prices of patented drugs and the fact that the general public in poor countries would not be able to afford essential lifesaving drugs at the current prices.[22] As the CEO of Cipla, Dr. Yousef Hamied, pointed out, "Should millions of Indians be denied the use of a life-saving drug just because the originator doesn't like the color of our skin?"[43]

This argument is about drugs available for not only domestic use, but also for global challenges, particularly for infectious diseases in Africa and elsewhere. The generic firms say that it is largely through them that drugs and vaccines are affordable for low-income communities, and it is their products that help the global health community make major gains in combatting diseases like HIV, malaria, TB, and other infections.[44] They also argue that it is because of the presence of generic firms, and the competition that they provide through attractive and affordable prices, that research-based pharmaceuticals are forced to lower their own prices in developing markets. This was particularly true in the early 2000s with Cipla's cheaper HIV drugs that not only created goodwill among the African population and leadership for Indian-manufactured drugs, but also made it possible for a greater number of patients to receive care. The Indian and Chinese generic manufacturers do own a significant share of the African pharmaceuticals market and their products are widely available, even in small pharmacies in remote areas on the continent. Cipla, for example, is the largest provider of antimalarials in Africa and its products are available in most malaria endemic countries.

Generic manufacturers also point out the bottom line. They argue that unlike research-based pharmaceutical companies, they are less concerned about the net profits and that their fight is not against research, but against the monopolization of life-saving commodities through the evergreening of patents.[24]

The generic pharmaceutical companies do not present a monolith. They vary in their size, scale of operation, and their global footprint. They also do not have the same internal quality control across the board. Their unified view on counterfeiting and substandard products is less clear than that of the research-based pharmaceutical companies.

Interestingly, just like their research-based counterparts, they also see themselves as victims. In my interviews with the senior leadership of generic pharmaceuticals (largely in Pakistan and some in India), there was a denial that the generic products across the board were of poor quality. This argument is not without merit, since out of hundreds, or perhaps thousands, of generic manufacturers, not all are complacent or complicit. A number of Indian manufacturers continue to be major players on the global stage and have strict quality assurance measures. Cipla makes nearly twenty-one billion tablets annually and has continued to maintain high standards of quality.[35] They export nearly 55 percent of their product, including to high-income countries.[45] Their international footprint is also increasing with new plants being built in Uganda and South Africa.[46] However, the emphasis on quality control is not the norm in all generic firms.

There is resistance, even among the top regulatory bodies in countries with a large pharmaceutical industry, to create standards that are similar to what exists in the European Union or the United States. The top Indian drug regulator, Mr. G. N. Singh, recently remarked that if he were to apply US standards in inspecting Indian facilities, he would have to shut down almost all of those.[47]

In Pakistan, the larger manufacturers such as Ferozesons want tighter regulations, and enforcement of better standards, much in the same vein as members of IFPMA. For Ferozesons, streamlining quality control would ensure fairer competition and would make it difficult for substandard manufacturers to operate.[48] However, smaller manufacturers, who make up the bulk of the market and are active in the Pakistan Pharmaceutical Manufacturers Association (PPMA), feel that regulation is already adequate, and any intent for additional regulation is driven by bureaucratic corruption and overreach.[49] The argument of PPMA members is that excessive regulation does little to improve drug quality, unless it is carried out with the necessary tools and the right intent. They point to large-scale government lapses and corruption and say that in the absence of accountability within government, excessive regulation would do more harm to their industry than good. This is despite cases where a large number of Pakistani-manufactured drugs, bound for Afghanistan, have been found to be counterfeit and substandard.[50]

In response to my questions to Pakistani drug manufacturers about how to ensure and regulate quality, I got varied answers. In general, there was no unified voice, and executives spoke about their respective firms and not about the generic pharmaceutical industry in general. This was strikingly different from their perspective on regulation and the regulatory authorities, where there was clear convergence in the opinion and unity in the position.

In general, there was also a strong sense of denial that quality was a serious issue in the local industry, or among generics in general. When I showed general statistics about quality being a global concern, some of the manufacturers pushed back and said that there had never been a quality issue in their own firms. It may or may not be true, because it is hard to attest that claim. Quality audits are rare and even if they do happen, the reports of such audits are not available to the general public or for research. Upon my insistence, including pointing to bad drugs manufactured in Pakistan sold in neighboring countries, as well as instances of deaths due to poor quality around the globe, the blame from generic manufacturers in Pakistan was pinned on a few bad apples in the industry that ruin the reputation of the industry as a whole.

The exact number of those bad apples, or whether those apples are temporarily bad or permanently rotten, no one could answer.

One thing was clear: There were many gulfs of trust. They were present between all the actors on this stage. There was distrust between the large research-based pharmaceuticals and the generic manufacturers, the generic manufacturers and the regulators were not on the same page, and the big multinational pharmaceutical companies and the regulators in low and middle income companies were distrustful of each other.

The public, in the meantime, is not convinced that the pharmaceutical companies, large or small, had their best interests at heart.[51]

8

The Disconnected Citizens

Paul Newton is a soft-spoken British doctor, who is based in Laos and has spent his career researching the issues of substandard and counterfeit medicines all over the world. For nearly two decades, he has been spearheading initiatives to test the quality of antimalarial drugs around the globe. He is often the first person to be consulted whenever there is a new story about the proliferation of poor-quality drugs. Paul has argued that the global achievements in reducing malaria-related deaths are in danger of a significant reversal if the problem of substandard and counterfeit drugs does not get under control.[1] This is particularly true in sub-Saharan Africa, where billions have been spent, and money continues to pour in, to create sustainable malaria-control programs. But the very sustainability of these programs is under threat, argues Dr. Newton. As resistance develops, the frontline of our arsenal, particularly artemisinin-based combination therapies (ACTs) will no longer be effective. In the region where Newton is based, in Southeast Asia, there are already reports of ACT resistance, driven in part by the widespread availability of poor-quality medicines.[2] Should this start to spread to large population centers in Asia,[3] or in sub-Saharan Africa, there would be significant impact both on public health and the success and sustainability of malaria-control programs.

Paul is also someone who wears many hats. He is simultaneously a physician and an advocate for more investment,[4] a policy expert often called upon at various international forums to share his expert opinion, and a leading epidemiologist and clinical science researcher in tropical diseases.

He is also one of the few people who have taken up the issue of estimating the real impact of substandard and counterfeit medicines, in terms of lives lost. While it may seem a basic problem that should have been resolved by now, in reality it is anything but straightforward.[5] Researchers, policy makers, and public health professionals need statistics and evidence, beyond anecdotes and personal observations, to make arguments for the growth or scale-back of programs or to highlight new challenges. Numbers are also needed to raise funds and increase awareness. But estimating the number of people who die due to poor-quality medicines is particularly difficult.

First is the basic issue of data on mortality. The majority of the deaths due to poor-quality medicine occur in low-income countries. In these places, the data collection is difficult to begin with and hardly the norm in local communities.[6] Simple mortality data, irrespective of the cause, itself can be hard to get as has been argued by several economists and public health researchers.[7] Even the data that do exist can be questionable and cannot be verified by independent and rigorous measures. Even something as basic as counting the number of people who live in a particular place is not without conflict and political interests. In Pakistan, for example, the most recent census was in 2017, and even before the results were announced, the provisional results started to deeply polarize the society. The last rigorous census, before the 2017 one, was conducted nearly three decades ago. Bizarre as it may seem, there are deep underlying political calculations putting off something as fundamental as a census.[8] Pakistan continues to be a feudally dominated society, and representation in parliament is based on the population in a given region. With urbanization and internal migration within the country, the demographics have changed and have become less favorable to the traditional feudal powers. A new census would automatically trigger redefining new boundaries for federal and provincial representation, and hence would challenge the old power. This chance for a potential change remains unacceptable to the old guard and the feudal lords, and they fiercely resist any new census.

While a census may have political facets, the lack of reporting on deaths also have cultural and religious angles. In many societies, particularly largely Muslim societies, the dead are buried immediately as part of religious and cultural customs. In rural areas, with little education and weak presence of any population bureaus, the deaths are not reported in time for updating any records. Additionally, death certificate issuance is a practice confined largely to urban areas. Thus official death numbers are not always trustworthy.

The concept of autopsy or postmortem is also a foreign concept in many societies, and the resistance to it is not just in the rural communities.[9] This complicates the problem further for various reasons.[10] First, there is little awareness among the general population to suspect any wrongdoing, and even if they were to suspect something, it would be next to impossible to get an autopsy. Second, a postmortem is an expensive pursuit. In poor communities, getting an autopsy done would be financially very challenging, perhaps even prohibitively expensive. With the government being complacent or not interested, the cash-strapped family interested in an autopsy would have to bear the costs. Also, not everyone who dies from poor-quality medicines dies in a hospital setting, and hence bringing the body back to the hospital for an autopsy, at a place that is inundated with patients, would be extremely difficult. Finally, the competence of the staff carrying out the autopsy is also not always assured. Because poor-quality medicines may leave different fingerprints than the typical cases of malpractice or crime, it would be challenging to clearly correlate the death with the consumption of a substandard drug, unless there was a clear case of poisoning.

It is important to note that poor-quality drugs are consumed for a variety of diseases and not just for the ailments for which they are prescribed for. For example, malaria drugs may also be used to reduce fever, in some communities, without finding out whether the cause of the fever is indeed malaria or not.[11] This problem is particularly acute with antibiotics, which are used widely and are available freely.[12] The practice of self-medication continues in a large number of low-income countries to this day where doctors overprescribe, and patients stock up on antibiotics, use them as they see fit, and rarely finish the entire course. Over-the-counter availability of nearly all medicines, including ones that should never be given without a prescription, and lack of awareness about resistance make this problem particularly severe. Thus, it becomes particularly difficult to correlate the consumption of a drug with an adverse outcome, when a wrong or a poor-quality drug is consumed. Did the death occur due to the poor quality of the drug, or did it occur because of the wrong choice?

Third, attributing a death directly to poor quality medicines is difficult because a patient's underlying health conditions may contribute to fatality as well. This is one of the biggest hurdles in both estimation of those who die due to poor-quality medicines and also in creating awareness among the family of the deceased. Because people who may be consuming poor-quality medicines may already be sick, and sometimes gravely sick, it is difficult to clearly correlate their passing with the consumption of a bad-quality drug. The presence of underlying conditions is also a disincentive for the family of the deceased to get alarmed, or to ask for any investigation, or request an autopsy. In these cases, it remains unclear whether the presence of a spurious drug was a real cause of death. Did it accelerate the decline of health, or was it a placebo that did not do anything and the patient died of complications unrelated to the drug?

Fourth, the effect of poor-quality medicine may not be immediate, but may contribute to worsening of the condition and the death may occur sometime later.[13] This makes the problem of estimation difficult because the substandard medicines may increase morbidity and worsening of the condition, but may not lead to the fatality right away. Poor-quality drugs may also cause adverse reactions, decrease immunity, or affect the normal functioning of the body, which, combined with other ailments, may lead to mortality. Therefore a direct connection, while certainly present, is nontrivial to ascertain in a strict legal and medical sense.

Last, but not least, substandard drugs may render good drugs ineffective.[14] The issue of resistance, in not just antibiotics but also antimalarials may lead to the ailment or infection that ultimately causes the death of the patient. A poor-quality antibiotic would create resistance against a good drug, and therefore it would make it much more difficult for the clinician or the healthcare provider to control infection with good-quality antibiotics. Given that the supply is often short and not all facilities in low-income settings have access to the whole arsenal of antibiotics, resistance against one or a few may severely affect the outcome for the patient.

It is, in part, due to the presence of these five challenges, individually and collectively, that good estimates are hard to obtain for measuring the impact of poor-quality medicines on overall mortality. However, despite this daunting task, there are statistical methods now being developed and utilized that may make the estimates more robust. Newton and his colleagues at Oxford, Laos, India, and at the Center for Disease Dynamics, Economics, and Policy in Washington, DC, set out to do just that.[15]

Right away the researchers knew that their task would not only be hard, but also that any number that they would get would not be exact; it would be an estimate of the problem. The team was fully aware of the challenges but also knew that a systematic study and the estimate resulting from robust methods were still a very meaningful outcome for researchers, public health practitioners, and policy makers.

To make the problem relatively manageable, the team, led by Dr. Ramanan Laxminarayan and including Dr. Newton, focused only on malaria. Malaria is something Newton and his colleagues at the Worldwide Antimalarial Resource Network are intimately familiar with, both in terms of the clinical dimensions of the disease and also in terms of the various therapies that are used.[16] Malaria is a problem that is particularly affected by poor-quality medicines, and Paul has had decades of experience with malaria medicine quality.

To drill down even further, the team of researchers focused only on children under five, again a group that is particularly affected by poor-quality medicines. Improving health outcomes in this group can make a huge difference in the outlook of public health in a country. Finally to define the geographic constraints of the problem the research team focused only on sub-Saharan Africa with its thirty-nine countries. Once again, this was a choice inspired by a heavy burden of disease and the reports suggesting the prevalence of poor-quality medicines in the region. So in essence, they started working on a tiny sliver of the whole problem, by confining age, geography, and disease. The argument was that if we could even estimate, with good measure, the impact of the problem in this particular region and with this particular demographic suffering from this particular disease, then we might be able to extrapolate this to other regions, or at least begin the process of better understanding the global impact.

As the team had expertise in a number of disciplines ranging from economics to data modeling, public health to microbiology, they were equipped to use rigorous tools to estimate the impact. The team used methods that incorporated several key components into a statistical model to predict the mortality rate based on key health parameters of the region of interest.[17] The key components of the model were the published reports of poor-quality medicines, household surveys on children under five who suffer from fevers, the fraction of fevers caused by malaria, the fraction of fevers that are treated with antimalarials, and private-sector estimates and demands of antimalarial sales. In other words, they took existing data on disease, death, and poor-quality medicines and from that data estimated how many children would have only received substandard medicines, if that was all that they were getting.

By design, this strategy was going to provide a lower-end estimate because of the problems of underreporting of death and disease. Given the circumstances, scarcity of any real and robust estimates and analysis, even underreporting was a major step forward. The team of researchers estimated, through their model and analysis, that nearly 122,350 children under five die annually because of poor-quality antimalarial drugs.

This number is staggering, given that this is only in sub-Saharan Africa, only for malaria and only for children under five. Nigeria, Uganda, and the Democratic Republic of Congo made up nearly 76 percent of all of these deaths, with the number in Nigeria alone reaching nearly 75,000.

While the statistics are sobering and depressing, and point to gross inefficiencies in the public health system, there are questions that go beyond statistics. It is perhaps easy to pin the blame on the government, and by no means should the government be absolved of their responsibility. There is some serious work that needs to be done by the government and regulatory bodies. But what are the citizens, who are most affected by this crisis, doing? Are they just on the receiving end of this crisis, helpless and passive recipients of the calamity? Are they complacent, too poor and too weak to force their governments to take action? Is the problem even on their radar and is there enough awareness to address the problem? And while technological innovation may not always be possible and beyond the reach of many who may not be technically savvy or trained, are the citizens concerned about the problem and engaged in creating solutions in any real way?

The answers to these questions are just as sobering as the statistics above.

While the estimate focused on sub-Saharan Africa, the problem, most certainly, is not limited to that area. Neither are the challenges that come from a disconnected citizenry. Citizens across the world, with a high burden of disease, poor adherence to prescriptions, and several serious incidents indicating the widespread presence of substandard drugs, seem to be poorly engaged in resolving this crisis.[18]

A recent study published in the journal *Lancet* points to a dangerous level of resistance against leading antimalarials, developing all the way from Thailand to the India-Myanmar border.[19] Poor-quality medicines continue to contribute to this challenge, leading to widespread mortality and morbidity in the region. With over a billion and a half people living South Asia, the issue of resistance at the doorstep of India, Bangladesh, and Pakistan is deeply concerning. That said, when compared to other issues, such as terrorism, crime, political corruption (associated with bribery, money laundering, or financial irregularities), environment, and other local problems, this problem does not make the news often enough. The research study that looked at malaria resistance in Southeast Asia and reaching the Indian subcontinent got a lot more press in countries outside the region than those in the epicenter or the vicinity. Why was that? Why are the citizens not bothered, and why do they not spring to action or demand better-quality drugs? And if there are some nascent efforts, why is that strategy not creating a chain reaction of other efforts or generating sustainable solutions?

The problem of the lukewarm enthusiasm among the citizens has multiple layers, all of which require serious examination. Within these layers, the first and the most direct connection between the consumer, supplier, and the drug starts with the pharmacist. It is important to understand that everyone who gets to work at a pharmacy in a low- or middle-income country to dispense drugs is not necessarily a qualified pharmacist.[20] In fact, many workers are just employees without any particular training. They might as well be working in any other retail industry. There are few, if any, laws that require that a pharmacist has to be present at the pharmacy at all times. Even if there are such laws, they are seldom enforced.

In situations where qualified pharmacists may even be present onsite, it would hardly resolve some of the deeper problems. This is due to gaps in training and capacity of those who become pharmacists.[21] The training of the pharmacists gives us some clues about their inability to fully grasp the problem.[22] In many parts of the developing world, pharmacy students are not directly exposed to the issue of substandard and counterfeit drugs. The curriculum requirements are highly inconsistent.[23] Curricula of pharmacy programs in low and middle income countries in Africa and Asia shows that the programs are highly inconsistent in their introduction to the topic of substandard and counterfeit medicines. The introduction and analysis of the topics ranges tremendously, and in several programs there is a discussion on the importance of pharmacovigilance but nothing comprehensive on the causes, manufacture, and proliferation of poor-quality medicines whatsoever.

Even within a single country there are tremendous differences in the curriculum. The curriculum approved by the Higher Education Commission in Pakistan has a theoretical component and a practical component.[24] The theoretical component talks about quality control in manufacture and talks about the telltale signs of poor-quality drugs, but it is archaic and lacks more recent global trends. It also talks about the problem in abstract ways disconnected from practical challenges. For example, it talks in general terms about what can go wrong during manufacturing, but fails to discuss the local challenges in Pakistan, issues with sampling and poor-quality medicines prevalent in rural areas, the gaps in policy and regulation, and technological limitations of existing tools. Other parts of the world, such as programs in Ghana, have curricula that do not even go through those basic details.[25] Introduction to the limits of technology and the emphasis on why newer methods are needed at the point of care, and at the point of need, are rarely discussed in most low and middle income country pharmacy programs.

The theoretical aspects of the program have severe limitations, but they pale in comparison to the much bigger problem in the practical part. The practical component, training in the lab and familiarity with various analytical methods, is often a requirement to complete the degree. Unfortunately, major gaps in this requirement leave a lot to be desired.

The practical component of the curriculum depends on the availability of lab equipment, training staff, and supplies. In a developing country, any or all of them may not be available at any given time.[26] There are few faculty members who are trained to address the problem, are aware of the latest technologies, or even have the capacity to teach what the problems in the field are. Their training may be outdated, and they have had few opportunities to stay up to date through conferences or symposia. The instrumentation available in lab facilities is often in a state of disrepair. The lab facilities are poorly managed. The instruments do not always work and the ones that do have logged more hours than the capacity of the machine. Without a clear mechanism to refurbish or buy new instruments, the labs continue to stay dysfunctional.[27] The problem of lab equipment not working in the desired manner is further complicated by the lack of lab supplies and resources that are needed to keep the laboratory functional.[28] Thus, even when the curriculum asks for the practical component and hands-on training, there is little capacity to carry out those activities in a reasonable and reliable manner. The consequence of these lapses in training is the gap between what the curriculum ought to do and what it ends up doing in terms of training the workforce that forms the bridge between the public and the drug supply.[29]

While it is quite likely that a pharmacist in a developing country is aware of the problem, it is highly unlikely that he or she may have fully understood the problem as part of the curriculum. It is more likely that the awareness stems from personal experiences and anecdotes. The rigorous grounding that is needed from a robust curriculum, analysis, and discussion is clearly lacking. Those who are being trained as potential regulators, distributors, store managers, or even researchers are never exposed to the problem in a comprehensive and rigorous way. With pharmacy being a bachelor's level degree in most countries, individual theses are either not part of the curriculum or in cases that they are, they lack novelty and rigor, and the projects are often repetitions of previous theses and prior projects. The ability of these individuals, who are coming out of national programs, and are supposed to be the leaders in national pharmaceutical efforts to contribute intellectually or practically to resolution of the problems, remains extremely limited.[29]

There is no single answer to the question of why, despite the widespread public health impact, the problem has not made it to the curriculum in a comprehensive manner. It is likely a combination of several factors.[30] The first is the lack of innovation in curriculum development. This lack of innovation is not just a problem in pharmacy or pharmaceutical sciences but a problem seen across the board in curricula in the postcolonial world.[31] Developing countries, by and large, have demonstrated inertia in creating innovative curricula in the sciences since independence from the colonial powers in the last five or six decades. The scope of the studies, and hence the curriculum, has either remained the same since colonial times or in recent years has been copied, in some cases verbatim,[32] from other developed nations, which have very different kinds of public health challenges. There is a sense that the developed countries have "made it" in science, and adopting their curriculum, will overall be a step in the right direction. In some cases, universities and researchers from the more established universities of higher-income countries have also contributed to the problem by promoting their own curricula, in some cases verbatim, as that is what they are most familiar with. This reliance on curricula from foreign institutions has led to a poor understanding of the local challenges, and it has stifled conversation on critical thinking.[33]

The curricula in medicine and pharmacy have also not fully embraced the socioeconomic aspects of the challenges.[20] The curricula in pharmacy are interdisciplinary to the extent that they include life sciences and chemistry, but not beyond that. Problems that are inherently interdisciplinary, and require interaction between social, biological, and regulatory sciences, are almost always ignored.[34] Departments of pharmacy rarely engage with social scientists, demographers, public health professionals, or public advocacy experts in any meaningful way. Academia in numerous low-income countries is a conglomeration of silos where interaction with outside disciplines is minimal, structurally separated, and often discouraged. Beyond technical writing, and a little bit of required social studies or history mandated by the ministries of education, there is little understanding of socioeconomics in the curricula of schools in English-speaking sub-Saharan Africa, India, Pakistan, and Bangladesh. An understanding of broader social issues is either never taught or if it is taught, it is discussed in a superficial way that is nothing but a burden on the student. Similarly, a detailed analysis of the regulatory framework is almost never taught in most places and remains the domain of legal scholarship and politics, topics that are beyond the interest or scope of pharmacy schools.

Connected to the lack of interdisciplinary activity is the lack of encouragement for innovation.[35] The curriculum is essentially set in stone and scripted for each term, and the students do not learn how they can contribute to solving the problem of counterfeit and substandard drugs in a truly multidisciplinary manner to impact the society at large. The problem of counterfeit, substandard, or poor-quality medicines, if discussed, is presented as a challenge driven in large part due to political corruption, outside interests, and criminal activities and is not a problem that can benefit from the intellectual contribution and innovation of pharmacy students, experts, and researchers. Stories from outside the academic sources, such as newspaper articles, focus on the corruption and sensational aspects of the problem, and do not connect the challenges as a problem that can be debated or discussed in rigorous manner in the classroom or research settings. This singular focus of the problem that is devoid of the technical component, which may be of interest to students, therefore provides little incentive for students and research scholars in pharmaceutical sciences to create robust solutions. This is further complicated by the fact that there is almost no opportunity to do an independent study or a project as part of the curriculum.[36] Therefore a rare student who is interested in developing a creative solution to the problem that he or she sees in the community would be unable to do so within the framework of the curriculum.

Innovation often requires a bridging of disciplines, and it is equally important for students, scholars, and researchers from other areas to engage with pharmacists and pharmaceutical scientists to address the problem. However, the problem of engagement or silos is not just limited to pharmaceutical science; it is also present in curricula of science and engineering and other disciplines that may interact with pharmaceutical sciences.[37]

This problem of a lack of interdisciplinary scholarship and out-of-the box thinking in engineering and science curricula is particularly acute in low- and middle-income countries.[38] Engineering curricula are almost bipolar. On the one hand, they represent the rigid structures created during the colonial times that reflected the course schedule, teaching methods, and educational strategies of the colonial powers, and on the other hand, the recent changes in curricula present a fascination with the most cutting-edge problems in the high-income nations. Problems that are inherently local, in engineering and health, are typically ignored in the curricula.[39] These local problems are considered less exciting and also harder to publish in scientific journals. As a result these problems are less likely to be addressed through teaching and research.[40]

This bipolarity extends in other dimensions as well and is out of step with the current global trends of training. For example, medicine and engineering are considered two polar opposites in the sciences that are never to be mixed. In Pakistan, after the national tenth-grade exam (called matriculation), students interested in science have to choose either a pre-medical curriculum or a pre-engineering curriculum, which allows them to apply to medical colleges or engineering schools.[41] If they choose a pre-medical curriculum, they opt for biology and cannot study mathematics. If they choose the pre-engineering curriculum, there is no option to study biology after the tenth grade. The national system of high school education in Pakistan does not allow a student to study both subjects simultaneously, and hence problems at the interface of biology and engineering are decidedly outside the domain of student inquiry. Recent developments are providing students interested in engineering to opt out of any biology whatsoever, by replacing it with computer science.[42] In the world that desperately needs problem solvers who can work with a multidisciplinary toolbox, and societies increasingly require systems thinking, curricula in developing countries often make the students unable to contribute to the long-standing and stubborn challenges of their surroundings.

The problem of awareness cannot be blamed on the pharmacists and their training alone. The next line of connection, after the pharmacists, goes through physicians and healthcare providers. Here again, the situation from the awareness angle is largely unsatisfactory. The problem of awareness among physicians in developing countries also suffers from major gaps, both in understanding the fundamentals, and in appreciation of the subtleties of the problem. In a number of developing countries, medical students never fully appreciate the medical significance of substandard drugs beyond the obvious cases of poisoning or adverse reactions. A more sophisticated understanding of immunity and resistance is lacking. This is driven in part, once again, by the deficiencies in the training curriculum of the physicians. Despite the fact that schools of medicine take some of the brightest students in a country, the curriculum often fails to provide them with the rigor needed to understand the problem of poor-quality medicines from a clinical standpoint. In Pakistan, medical students never take a dedicated course in immunology or cell biology,[43] making them unaware of why substandard drugs pose both short- and long-term challenges, and if the problem remains unchecked, they cannot look at how these challenges multiply across the society as a whole. Lacking something as basic as training in immunology makes clinical research by doctors on issues of resistance challenging, and the cycle of poor understanding and an unaware clinical community continues to operate. Resolution to this problem should be, in principle, straightforward through updating and modernizing the curriculum. Yet the curriculum update in general is centralized and too bureaucratic to reflect current trends, along with a strong resistance for any changes.[44]

While there is a clear link between increased mortality and morbidity due to substandard medicines, and there is evidence connecting resistance and poor-quality medicines, in general the medical community in most developing countries has failed to recognize this as a problem that is directly relevant to their profession or worthy of teaching in the classroom. As a result, doctors, who are a more powerful lobby than the average citizens, or even pharmacists, are unable to play a more constructive role in raising awareness in the society about the problem.[45]

While it may seem counterintuitive, the doctors themselves may also contribute to the problem. There are three main facets by which doctors and clinicians in developing countries make the problem worse. First, they are fundamentally unaware of the mechanism of drug action and not up to date on resistance literature.[46] Staying current with recent literature is not the norm, as the publications may be expensive, and there is not a culture of staying engaged with clinical literature among doctors working at the primary or district health facilities. As a result of this lack of knowledge, they do not take a strong stance on the issue and are likely to prescribe drugs even when there is evidence of lack of therapeutic efficacy. This, falls in the category of poor training, ignorance, complacency, and negligence.

Second, doctors who are also unaware of the problem sometimes can promote practices, even in good faith, that are counterproductive and at times outright dangerous. The issue of quality and efficacy of oxytocin, a drug used to control postpartum hemorrhage, is well known in the literature and has been highlighted in various international meetings and symposia.[47] Oxytocin can make a difference between life and death, quite literally. Its proper use, in a proper concentration, can save a woman from one of the most common causes of maternal mortality. Yet the quality of oxytocin has been an issue in developing countries. In India and Nepal, when realizing that their oxytocin is no longer working on the patient, the doctors continue to give additional doses, until it starts to work.[48] In doing so, they are doing all that they can to save the life of the patient, but also creating a precedence of the bad practice of doubling, or quadrupling, the dose of the drug from what is clinically prescribed. This has led to the perception among some nurses and clinicians that women in South Asia require more oxytocin than women in other parts. In reality, it is not genetics, but the poor quality that demands additional doses. In a situation when oxytocin may be potent, the precedence of giving additional doses can be toxic and dangerous.

The third facet is along the lines of corruption. Aggressive marketing by pharmaceutical sales agents, who work on commission, and a lack of rules that requires doctors to declare their income or honoraria from pharmaceutical companies lead to the clinicians accepting funds from companies, in return for prescribing their drugs, even when the drug may not be most efficacious.[49,50]

This leads to not only broad-based resistance issues, but also poor health management. Similarly, in the absence of quality control mechanisms, aggressive prescriptions increase sales and demand. The increase in demand, without quality control, can further expand the issue of poor-quality medicines. Without a mechanism to hold doctors accountable, or even question their judgment, in societies that put doctors on a high pedestal, the actions and the inactions of doctors go largely unnoticed and unpunished.

Doctors in training or in practice, pharmacists, and students in general cannot be blamed alone for lackluster engagement on their part. The burden is also shared by NGOs and industry organizations that, on one hand, make awareness of the issue a central theme and, on the other, miss the opportunity to fully engage local citizenry. While seminars on the issue of quality control are often sponsored by the global players, they tend to attract the same group of people who tend to give the same presentations, and the engagement with citizens, in particular students, is completely lacking. There is a fundamental disconnect between academia, research, and policy in many developing nations.

In 2014–2015, for example, the International Federation of Pharmaceutical Manufacturers Association (IFPMA) conducted several international conferences in various parts of the world. I happened to attend the one in Senegal in April 2015.[51] These conferences were focused on strengthening regulation and had attracted local regulatory authorities, some public health professionals, members of pharmaceutical industry, domestic and international NGOs, and members of various international bodies ranging from the World Health Organization (WHO) to the World Bank. In some cases, government officials from the Ministry of Health also attend these meetings. The IFPMA seminars were all on target and the message resonated with those who were present. Yet something was amiss. The meetings focused largely on regulation and capacity building, and very little attention was paid to education, training, research, and student engagement.

There were no student presentations or research poster sessions to highlight current local challenges, and there were not any platforms to motivate the local students to innovate. It was a missed opportunity. There was also no discussion of how to engage researchers and scholars from the local universities and colleges, or how to incentivize local ownership of the problem.

The IFPMA assembly in New York in 2014 that attracted high-profile industry leaders was marginally better with a session on young innovators.[52] There were also a couple of students among hundreds of attendees present, largely because they happened to be working in areas of public health that touched on the issue. There was, however, once again little discussion of creating local owners of the problem, and creating incentives for students in developing countries to engage, addressing local gaps in curricula and building sustainable coalitions. This is a problem that is well within reach of IFPMA and other organizations and can be addressed by engaging local students to create awareness and interest.

The problem of disconnected citizenry is also driven in part by how the story is portrayed in the media. It would not be fair to argue that there is no media campaign in any developing country. There have been stories reported, in print, electronic, and more recently social media that have generated some action, even in low- and middle-income countries.[53,54] The electronic media tend to play the most important part in the developing countries, because print-media stories are less likely to create awareness as citizens may not be literate or engage with print media in the same manner as they would with electronic media. Social media also favor the literate and savvy among the population, which in many parts of the developing world, would miss the rural and uneducated communities.

In order to attract the attention of the viewers the media stories are typically sensational—with a flavor of uncovering a big coup or a scandal. The sensational story is largely about a bad guy making a fake pill in a night operation, or a major international mafia ring creating millions of knock-offs. While not inaccurate, the stories about the lack of regulation, or the mislabeled products or products that degrade over time, are never told in earnest. The stories are also unlikely to talk to experts and discuss the multiple dimensions of the problem, including the lack of appropriate technology or the widespread system-level issues, such as lack of quality control during transportation, loss of electricity in storage, and poor training of pharmacy staff. The nuances that are associated with the problem are never communicated with the clarity that they deserve.

The media houses, in their defense, argue that in the absence of having a culprit that needs to be shamed, or sensationalism that creates strong viewership, there is no viewer appetite.[55] They also point to a limited time slot, of a few minutes at best, that is available to cover the story. The nuances would take too long to cover, they argue. The state-owned media, on the other hand, are either largely unpopular with limited viewership or unlikely to talk about challenges that put the blame on the government for its inaction.

In an era of social media and instant news, the likelihood of the news sticking in our memory is also low and is unable to generate traction. Also, as noted above, pinning death directly to a bad drug is often challenging and requires extensive analysis, so a story without a casualty number is less likely to be attractive to the media. That said, some in the print media in recent years have carried out several impressive and detailed stories, though largely in the United States and Europe. The real impact of these stories in creating real change on the ground, or providing researchers a sputnik moment, or charging the citizenry to demand action remains very hard to quantify. Stories by the *Wall Street Journal* about the prevalence of counterfeit and substandard antimalarials in Angola[56] and by the *New York Times* about the contaminated cold medicine in Panama[57] have resulted in a broader discussion of the issue. However, both the *Wall Street Journal* and the *New York Times* have significant resources at their disposal to thoroughly investigate the issue, get the facts on the ground, and interview researchers and stakeholders in multiple countries. These resources are often absent from the print media in low- and middle-income countries that are finding it harder to stay afloat in the era of social media and sensationalism on television.

The framing of the problem, as a sensational story, has another fundamental disadvantage. It takes the problem out of the hands of ordinary citizens, students, and researchers. The problem is presented more as a problem of policing and crime than something that an engineering student or a scientist can address. By not presenting it as a scientific challenge that has a social dimension, a political problem that requires a new innovation, or a public health challenge that desperately requires local solutions, the ability to engage an inspired scientist or engineer is lost, and so is the ability to create a sustainable, robust, and affordable solution.

Empowering citizens requires awareness, and some governments are taking steps to create this awareness by providing citizens with tools to engage. Some campaigns, in the era of smartphones, have also tried to engage citizens by asking them to inform the government via various apps. In Pakistan, this initiative has been rolled out in the Punjab province.[58] The problem of the culture, however, remains. First, while smartphones are gaining popularity, the appeal of the various apps, and their full potential, has been limited to educated citizenry in urban areas. In places where the problem is most acute, in rural areas, with a sizable number of uneducated citizenry, the uses of smartphone for surveillance and reporting of drug quality are marginal. There is also a fundamental danger in overselling the solution. Smartphones are being used as substitutes for system monitoring and surveillance, and governments are relying on them to hide their inefficiencies.[59] A smartphone may be able to capture some information but does not substitute for testing the quality of a pill or perform a rigorous analytical test. What is needed is a recognition of the potential power, and fundamental limitations of smartphones, and the engagement of citizens not through gadgets alone but through better awareness.

Some countries, such as Nigeria,[60] Kenya,[61] and Indonesia,[62] have tried to create hotlines where citizens are encouraged to call in should they come across suspicious drugs. The toll-free number is designed to create a bridge between those who experience something suspicious and the regulators. In principle, this is a useful idea but unfortunately suffers from the same logistical challenges and bureaucratic hurdles as many other good ideas.

First, the concept of calling in, particularly to a service that may be in a distant town, upon seeing something suspicious is not rooted in the local culture. With widespread distrust of government, people are unlikely to call in and talk about suspicious drugs on the phone with someone they do not know. With a significant trust deficit, and lack of action, there may be little reason to call a hotline and talk to someone in another city. This deficit stems from the feeling that government is inactive, the solution is only a lip service and any real requests would only be acted upon after substantial delays, which further fuels distrust. The vicious cycle makes it hard to implement this policy.

The government also does not share what action, if any, has been taken.[63] The results, if available, are only available in the form of policy documents or papers on the government websites, if at all. This may be useful information for researchers and policy makers, but it does little to engage citizens who are not used to reading policy papers or scouring data from websites.

Groups of citizens in developing countries, as doctors and teachers, as investors and policy analysts, do fall short in carrying out their responsibility to address the problem in their respective spheres. But so do others, who may live in more economically stable nations, in creating avenues of true global awareness, engagement and training, and understanding both the culture and the context needed to address the problem.[64] This problem may have hot spots in certain parts of the world, but it remains a truly global problem, and global citizens who may have intellectual, financial, and political resources remain largely disconnected in understanding, owning, and solving the problem.

9

The Technological Fix?

Behind every major public hospital that I have visited in a low-income country, be it in Pakistan or in Papua New Guinea, in Nepal or in Namibia, there is a graveyard. It is not the graveyard where bodies of the deceased are buried. It is where old equipment, often donated, is left to rot, in public. This technology graveyard may be close to the wards, or half a kilometer away, but the stacks of hardware are there for anyone who is curious enough to inquire. Monitors and pumps, defibrillators and cPAP machines: all are there in heaps and piles.[1] There is no decomposition of organic matter in these graveyards; it is made up of rusting metal and of plastic that may stay in the same shape and form for millennia.

It is estimated that nearly 80 percent of the hospital equipment and hardware in low-income countries are donated.[2] Some studies suggest that about 40 percent of donated equipment is nonfunctional, whereas other reports put that number to as high as 80 percent.[3] Regardless of the exact number, the fact remains that technology, both in terms of acquisition of the right technology, and the ability to create appropriate solutions in low-income settings, remain a major challenge.

The issue is not just with medical equipment at hospitals in developing nations. It is true elsewhere in the health system in these countries as well. During my trip to Senegal in 2015, I was told that at the national drug-testing lab, there are two high-performance liquid chromatography (HPLC) machines that are supposed to do drug testing and help regulators improve drug quality in the country. Only one was working and the other one had been offline for as long as they could remember. One of the staff members told me that in his years of working at the lab, he had never seen it work.

Is our failure to comprehensively combat the growth of substandard and counterfeit drugs due to our lack of having the right tools, at the right place and in the hands of the right people?

Those who favor only an economic and political solution to the problem argue that there are tools used by high-income countries that seem to be effective and robust. So does this mean that the technology problem has been resolved? Or do we actually need new tools, optimized for the challenges of remote places and resource-limited settings, that are context-specific, appropriate for the environment, able to work in the hands of the local workforce, and cognizant of technological, financial, and social challenges?

It is important to explore the question of technology in the context of drug-quality regulation and understand why after decades of effort, there are still major technological gaps for drug testing and efficient enforcement of quality standards.

Two basic arguments can be made here. The first one is that the issue is basically of governance, transparency, and policy.[4] This argument suggests that there may be a technological need, but the real issue is that of improvement in governance and policy and that existing technology may be able to resolve the problem, when combined with good governance.[5] This argument puts corruption, at the local, national, and even international level, as the central theme of improving drug quality. I have discussed the challenges of governance in the previous chapters. While better governance is indeed needed, this chapter explores the limitations of current technologies and the need for technological advancement in the fight against poor-quality medicines.

We start with exploring the hypothetical argument that new technology is needed to address the problem of proliferation of substandard and counterfeit drugs. We assume, for the sake of argument, that these new tools are needed most urgently in places where the need is most acute. We will test this hypothesis and analyze whether the technological need is real, or whether the technology issue been resolved already.

Starting with the question about whether new technology is needed, we will further explore a number of subsequent questions. For example, if it is needed, who will develop the technology? And why would that individual or institution do so? In other words, assuming that the company or individual will need some kind of return on their invention, what is the incentive to create something exclusively for places where the ability to pay for services may not exist? Would it lead to a viable business? Or in this case, will the innovation have to come only with a nonprofit entity in mind? There are questions that come up in technology development circles routinely. The most central one is about the scale-up of technology. How much would the local culture, customs, and regulatory structure matter in the scale-up? Does Pakistan need different technology than Peru where the needs may be different than those in Papua New Guinea? If so, would the cost of precision and accuracy be too high? Finally, we will explore the issue of new technologies coming up on the horizon, their potential in addressing the problem, and the barriers to scale-up and impact.

There is little debate that technology in some form is needed to address the issue.[6] At some point, testing must happen, and pure goodwill of the manufacturers, importers, and suppliers is not enough. Invariably, all approaches to test drug quality, whether at the central government level, or at the rural level, require some kind of technological solution. Whether it is a high-end mass spectrometer that may cost upward of a million dollars in just capital equipment and requires highly trained professionals, or the gold standard of HPLC (that costs over a hundred thousand dollars) or a simple paper strip that changes color upon coming in contact with the drug solution, some kind of technology is needed. Some of these instruments are commercially available, albeit with different prices in different parts of the world, while others are in various stages of development and field testing. Simple observations about packaging, and making sure that the box has not been tampered, with are no longer sufficient to comprehensively determine the authenticity and quality of the drug.[7]

The first question to explore is whether what we already have available on the market is sufficient or not. Here, there are two issues that need to be analyzed. First, there is the issue of viability, as well as efficient and effective usage. Is the technology that is commercially available, and used routinely in high-income countries, indeed viable in the developing nations? If so, why isn't it used as efficiently as it is meant to? Is that also a technological limitation issue or an interlinked issue of capacity to use and maintain?

Second, if that is not the case, and the current technology used by the Centers for Disease Control and Prevention in the United States, or the drug regulators working for the European Union is not appropriate for usage in Malawi or Honduras, is that due to the limitation of the technology? Or is that due to some other sociocultural or economic factors that require new technology that is relevant for the needs of the citizens of Malawi and Honduras?

To illustrate the first question, let us start with an example. The most commonly used and recommended gold-standard equipment by regulators is HPLC. HPLC is a relatively old technique that is based on the separation of various components within the sample. While separation based on size has been known for over a hundred years, some of the earliest work on HPLC was done in the early 1940s.[8] However, it was not until the late 1960s that the modern HPLC instrument found its way into the chemistry labs.[9] The way a typical experiment would work is that a sample, for example a tablet, that has been dissolved in a liquid would pass under pressure through a column. Various components in the solvent would interact differently with the column material, and as a result these components would flow differently. As they would elute differently, the concentration of each component can be measured with high precision.

Today, an HPLC instrument, columns, and various other components make up several billion dollars in annual sales.[10] HPLC instruments from brand-new ones to those that have already logged thousands of hours are part of drug-testing labs. HPLC, due to its robustness and performance, is the most common instrument used by regulators and pharmacopeia across the world to establish quality.[11] As this technology has been around for decades, nearly all countries have access to at least one or two of these in the central facility. Or at least on paper they do.

In terms of cost, it comes with a price tag of somewhere in the range of a hundred thousand US dollars and requires supplies, consumables, and reliable grid power. Additionally, it also requires trained staff with some reasonable familiarity with analytical chemistry and instrumentation. The equipment is also not portable. It requires a working lab, a stable and clean workspace, and ideally temperature control maintained in the lab environment. All of these elements add further costs in terms of maintenance and upkeep. Because it is not portable, it cannot be used at the frontline by a drug inspector during a sampling visit. The process of drug testing would require sample collection and then the shipping of those samples to the HPLC lab for analysis of quality and subsequent action or policy changes.

While HPLC is a workhorse in national drug-testing labs, there are serious challenges associated with its functioning. First, the cost is relatively high; as a result there are only a handful of these technologies available for testing. In Ghana, the number in the national Food and Drug Administration (FDA) lab is less than five[12] and in Pakistan,[13] the sixth largest country by population, the number in the national labs is also in low double digits. For comparison, there are several of these instruments in my university (in Boston) alone.

Then there is the cost of running the instrument. Let us assume that the instrument is used every working day (~250 days per year) and is able to test four different samples per day. A typical sample would be an aliquot from multiple tablets crushed together. This assumption is on the higher side and assumes that there is nonstop testing, because preparation and running the instrument for one sample takes about two hours. For a country with tens of millions of people, that require regular drug testing, this is far from sufficient. Even if this assumption of 250 days and four tests per day was to hold, and that everything ran smoothly and without a hitch, and that there was no dead time for the instrument being offline or being maintained, a total of a thousand samples would be tested per year. Even in this ideal, and perhaps unrealistic, scenario, the cost of having supervision of the instrument and consumables (which can be fairly expensive) would make it a high cost endeavor. In reality, the instrument does need routine maintenance; the staff does not work every minute of every working day, and the consumables require a tight supply chain and often take several weeks to arrive after an order has been placed. The problem is further complicated by the fact that at any given time, a health system may need regular testing in pre- and postmarket surveillance or may have to carry out extensive tests in response to an emergency and conduct additional tests for legal proceedings. The demand for testing therefore is continuous and in the absence of several instruments working full time, a backlog is quickly created and starts to slow down the entire system.

This snowball effect of the backlog, and a lack of personnel to maintain equipment, is quite common and significantly affects the overall performance of the lab.[14] The workload on the instrument also has a feedback effect, as it starts to influence and affect procurement. A new shipment of medicines is often accepted without any testing, because of the instrument having a long queue built up already.[15] Since the supplier often does not have the time or patience to wait for months (or years) to get paid before testing confirms the quality of the shipment, the buying country ends up procuring without any substantive testing. It is largely because testing at the national drug-testing labs cannot be done in a reasonable time period.

Finally, the issue of instrument maintenance cannot be ignored. Both preventative maintenance and regular upkeep are fundamental components of any hardware use, even in countries with a strong engineering culture and appreciation for the needs of technology. At the drug-testing labs, across the developing world, instruments often function below their maximum capacity due to a lack of technical expertise needed for preventative maintenance and are routinely taken offline due to problems that require external expertise.[16] This leads to two other challenges. Instruments that are donated may not have service contracts, and when they stop functioning they become useless. On the other hand, the service contracts themselves come with a hefty price tag and add an additional element of financial burden. With instruments that may be beyond the capacity of local technicians, the funding required for bringing the instrument back online becomes a hindrance. This is essentially a similar challenge to the ones faced by hospitals that end up creating hardware graveyards.[17]

Now let us turn our attention to the second big question—is there something fundamentally wrong with HPLC? Most certainly not. However, the fact that the health and regulatory system is unable to fully benefit from an existing technology is nonetheless a problem. So where does that problem stem from? Does it originate from the limitation of the technology in a resource-limited setting, or does it stem from the lack of infrastructure (human, capital and physical) that is needed for smooth running of the instrument? It is important to note that HPLC itself is not a technology at the point of care or the point of need, and therefore a fully functioning system would require multiple checkpoints, where technologies or inspections at the point of need would function in tandem with regular testing that needs HPLC. Thus, there needs to be samples and information that go into HPLC, and in the absence of the rest of the system working properly, the utility of something even as useful as HPLC would be minimal.

Similar arguments could be made about other high-end, precise, and accurate technologies, such as mass spectrometers, and their limited impact in the developing nations. While one can argue that the cost per test is actually not very high, when compared against the fact that the information attained is of the highest quality, yet the investment needed to reach that low cost per test, in terms of infrastructure and human capacity it is substantially high.

The need for rapid and point-of-care instrumentation has been felt for some time.[6] In the light of this need, the World Health Organization has recommended the use of a point-of-care testing system called MiniLab.[18] Developed by the Global Pharma Health Fund (GPHF), based in Germany and supported by the pharmaceutical company Merck, the MiniLab is based on the idea of a lab in a suitcase. It is designed to be self-sufficient, requiring minimal expertise, and has a detailed manual with a monograph on each drug it is capable to test. The technology at the core of MiniLab is thin layer chromatography, or TLC. TLC has been used for nearly a hundred years to separate components in a mixture. The idea is based on using a thin layer of silica gel or cellulose coated on plastic or glass, called the stationary phase, as a medium to separate the components of the mixture. The sample, in this case the solution containing the drug, is then applied to this stationary phase. The solution is called the mobile phase. Because of the chemical composition of the thin layer, different components of the drug mixture (i.e., the mobile phase) move up the thin layer at different rates. As they move with different rates, they start to separate. This separation can be visualized at the end of the experiment through various imaging techniques, such as exposure to UV light using a hand-held UV lamp.

While the test itself is straightforward, it is still not exactly point of care or point of need. The test does not require any substantial hardware or grid power; however, the number of key supplies needed to run every test makes it difficult to carry out in a short time at any point of care. As a result of the extensive number of experimental components needed, the MiniLab suitcase is nearly one hundred pounds in weight, making it difficult to carry around or operate at various points in the supply chain. Perhaps the bigger challenges stem from the limitations in quantification,[19] ability to test injectables and liquids, or the capacity to provide any information on dissolution that are often needed for regulators. Nonetheless, in a number of scenarios, where nothing else can provide useful information, MiniLab has provided highly valuable information to the regulators.[20]

In addition to the high-end mass spectrometers, the gold-standard HPLC, and MiniLab, there are a few additional optical technologies that are available in the market. These optical technologies use light in a hand-held instrument and do not need the pill to be destroyed. They resemble hand-held scanning guns seen at grocery stores or speed guns used by traffic police to monitor traffic. The underlying science is built upon spectroscopy, either Infrared or Raman. The instrument shines laser light on the sample, which upon absorption or scatter is collected and analyzed within the instrument with the help of a camera or a detector. The process is fast, nondestructive, and requires minimal sample prep. While these instruments are being used by pharmaceutical companies interested in the issues of brand protection and identifying true counterfeits, their penetration among the ministries of health and drug-regulatory sector is minimal.[21] This is due to several reasons. The first reason is the high cost of the instrument. Instruments can cost upwards of thirty thousand US dollars per device, with additional costs for the screening libraries that store the information of the reference standards. Second, not all drugs can be screened due to the limitations of the instrument and the range of signals that are detectable. This is particularly true for drugs that have multiple active ingredients (e.g., fixed-dose combinations) or drugs that give off a fluorescent signature upon exposure to light. The application for substandard drugs, which may have some active ingredient, and are not outright fake, also pose problems for these hand-held spectroscopic instruments. The application of these instruments has largely been for qualitative studies, and countrywide scaling up has been very limited.

An analysis of all three main categories of available instruments, from mass and liquid chromatography to thin-layer methods to hand-held spectroscopic instruments, demonstrates that the need for new appropriate technologies comes from a combination of instrumentation gaps and from the fundamental limitations of existing infrastructure. In the absence of development as a whole, and particularly development in the technical sector that is able to innovate, there will remain a gap that will have to be filled with appropriate technology. This supports our initial hypothesis that there is a need for new and innovative technological approaches; however, it leads to a number of additional questions about product development and sustainability that need to be answered.

First, we should ask what should the technology do, and what information should it provide to the users? While from a distance, the question may be obvious and identical in all situations, a closer inspection reveals that various questions require very different kind of technological solutions. The need for counterfeit detection is different from that of substandard drugs and different from testing or looking for degraded products. Each requires a solution that is both context-specific in terms of the geographic location and the nature of the product.

Let us first examine the case of counterfeits. Imagine a company X, that makes a particular drug Y. The company has the license to function and operate in a given country. Now, another company, let us call it Q, without having the license or authority, starts to make the same drug Y and starts to sell it to the consumers under the same brand as company X. They may offer the same or a lower price as what company X offers. Their quality may be the same or different. However, the issue here is neither of price nor of quality. It is in fact a legal issue of authority to make (and sell) product Y with that brand name.

The goal of our desired technology here would be to catch the product made by company Q, even if it looks identical to the product made by company X. A technology solution here may care less about the performance of the drug, but more about its lineage. The question that the user would ask, in this scenario, is not whether the drug is of the desired quality, but instead is the drug made by company X or company Q? Thus the technology does not need to look inside the package and identify the amount of active ingredient, but instead in some manner it should aim to determine the origin of the drug. This information, as one can imagine, is of high value to company X to maintain the integrity of their brand and to protect their profits.

Sproxil, a company based in Cambridge, Massachusetts, aims to tackle this problem through scratch-off barcodes and mobile phones.[22] The idea of barcoding is that good companies, operating with proper licenses, would print an alphanumeric barcode on the box or the packet of the drug. Only the company that made the drug would know the authentic barcode. The user upon receiving the drug would scratch off the barcode and call a toll-free number. The user would enter the number on the barcode when prompted by the toll-free number. The toll-free number would be connected to a computer or a service that would compare it against the database, find whether it is indeed authentic or not, and send a text message back to the user with that information. The idea here is to empower users to make sure that they have good stuff and not counterfeit products.

An effort like this requires partnership from three sources: the pharmaceutical company that is willing to put the barcode on its package, the user who is interested in making the phone call to find whether his or her product is authentic, and the phone company that provides the call in service and connects to the database. Sproxil and other companies providing this kind of service have had success in generating interest among pharmaceutical companies to provide barcodes on their packages.[23] This has also had a positive impact on generating awareness against counterfeit medicines. However, counterfeiters may set up their own fake phone lines and an unsuspecting consumer may not be able to tell the difference. Additionally, the burden to check the quality would fall on the consumer, and the method would not necessarily empower the health regulatory authorities. Furthermore, this method has limited applications in catching poorly manufactured products or drugs that degrade in the system due to storage and transport.

Now assume that company X, making drug Y, with a license to operate and all intellectual property intact, starts to make a poor-quality version of drug Y. This might be due to intentional or unintentional reasons. There is no breach of license, or that of intellectual property. The manufacturing is all in-house, at the facility owned and operated by company X. In this scenario, checking the packaging or the box would not inform the user or the regulator of the quality of drug Y. Pakistan's recent public health crisis in Lahore at the Punjab Institute of Cardiology was not due to a counterfeit drug. Instead, it was Eforze Chemicals, a company with all its licenses intact and working under the purview of the law, making a drug that was of poor quality and had failed to meet the quality standards.[24] Companies, even reputable ones, face quality issues, and every now and then some of the drugs made by these bona fide companies make it to the market. A user calling to find out whether the drug was indeed made by the said company would get an affirmative response, and the approach used in the scenario above, using a scratch code, would not be of much use.

Identifying poor-quality medicines in this scenario would require a technology that probes the active ingredient of the drug and its concentration. MiniLab aims to tackle this issue but has some key challenges of portability. Approaches that are similar to MiniLab, but more portable and easy to use, are being developed by researchers at the University of Notre Dame in Indiana.[25,26] The idea of these portable approaches is to have paper strips that are easy to use (like litmus paper) and change color upon interaction with the drug. Good drugs give a result in one color and bad ones result in another; therefore, the interpretation of the result is straightforward. However, there are further subtleties that need to be thought through for these systems to be used at various points in the drug supply chain. For example, these paper strips are not quantitative and may not be able to tell exactly how much active ingredient is present. But is that sufficient for regulatory purposes? What if the drug has some amount of active ingredient, but not the entire amount? Allowing drugs with subtherapeutic doses can increase antimicrobial resistance, and hence these paper strips can work at the frontline and raise red flags that could be further investigated.

Another problem that is often discussed by regulators and pharmaceutical scientists,[27] but ignored by the media and frontline healthcare workers is that of drug dissolution, which is fundamental to drug action and performance.[28] Every pill operates not only because there is an active ingredient of known amount present in it, but also because that active ingredient is released into the body. Imagine if drug Y is taken by the patient, but the active ingredient is never released in the body? Instead, drug Y sits in the gut of the patient like a little inert pallet that is eventually discarded by the body. Or what if the active ingredient is released too slowly, over a period of days, instead of minutes and hours? Or what if all of it is released immediately, making the drug toxic? Each drug that appears on the market is therefore required to have not just the right amount of the active ingredient but also has a profile by which the active ingredient should be released in vivo. The efficacy, and the quality, of the drug therefore rests on having both the right amount and the right release rate. Testing drugs for dissolution or release rate is a nontrivial problem and the technological challenges there would be different and require a multipronged approach. While there is a specialized instrument available for dissolution testing, it is expensive and requires a controlled environment and well-trained staff. As a result, most portable technologies sidestep the problem completely, as do lots of regulators in low-income countries.

The sheer volume of the drugs on the market and the time it takes to run each sample must be taken into account. Technologies like HPLC that focus on separation are able to test essentially any drug available on the market, but the time to test each drug is not in seconds or minutes, but in hours. On the other hand, technologies that do not use separation-based methods, such as Raman- or Infrared-based technologies, as well as paper strips, are much faster but are limited in the number of drugs they can test. There is a tension between the time it takes to test a sample versus the breadth of the drugs that can be tested.

Connected to the issue of technology, breadth, and time is an all-important question about the user. Broadly speaking, who is the user of the technology that is needed in the field? The needs of the consumer, ready to take the pill or give the syrup to the young child, are different from that of a regulator. The consumer wants a quick answer, in an easy-to-use manner, and perhaps does not care about all the details of the assay. Technologies such as the scratch-off number on the packet cater to this need and are geared toward the end of the supply chain. These are meant to empower the user. On the other hand, a pharmacist in a hospital, checking the quality of his or her drugs, may have different needs and may require more information than a simple yes or a no. With the high volume of drugs to be dealt with, he or she may also be less inclined to call a number on each batch. The regulators and members in the ministries of health have completely different needs altogether. They may be willing to forgo that a technology may not provide an answer instantaneously, but they may be more interested in finding more information about drug ingredients, presence of spurious elements, and even dissolution. Even within the regulatory bodies, the lab scientists may have different goals in mind than the inspectors on the street. A customs officer at the port of entry will also have different things that he or she may want to test, when compared to their colleagues in the regulatory sector. A pharmaceutical company may have completely different needs altogether that may be different than what other stakeholders express as their key priorities.

Similar arguments about the need of specific technology can be made about the nature of the medicine (solid or liquid, oral or injectable, etc.) as well as the point of testing. Should medicines be tested at the point of entry? Is it really necessary? Or should they be tested only right before the consumption by the patient? Can there be a single technology that can allow for both and for the stages in between? All of these questions are relevant, and it is precisely the complexity that arises from these questions that makes technology both extremely relevant and often limited in its scope to answer these questions.

There is a clear gap between needs in various parts of the supply chain in the developing countries, and technologies that are available to address those needs. This is driven by a combination of cost, limited local infrastructure, and challenges in human resources to maintain technology working at its full potential. This gap, and the sensitivity to the local needs, has spurred the development of new technologies.

The list of new technologies that are in various stages of development, testing, and deployment is relatively small. The technologies that are appearing on the horizon, aimed at addressing the issue of substandard and counterfeit drugs, can be divided into four broad categories. The first category is based on mobile phone–based systems that largely address the issue of counterfeit drugs.[29] The emergence of mHealth approaches, combined with an increase in the number of mobile phones and users across the world, has opened new possibilities for disease detection, management, and data collection. The mHealth sector has also seen new opportunities in drug quality control. The use

FIGURE 9.1.

Colombian authorities seized old vials being washed and reused to hold counterfeit drugs. Several technologies focus on checking the label to determine authenticity, and genuine labels can be difficult to make. By using authentic packaging and bottles, counterfeiters are able to go undetected. Reproduced with permission from Pfizer.

of texting and SMS, and in some cases images and videos, have enabled innovators to harness these platforms for drug testing. Technologies such as the ones developed by Sproxil and other companies, including mPedigree[30] in Ghana and ProCheck[31] in Pakistan, represent this area of innovation that look at the packaging to ensure that the product was made by a legitimate and licensed manufacturer.

The second category of technologies use paper-based systems to identify the presence of the active pharmaceutical ingredient (API), or a particular impurity, within the drug. The idea here is that the paper strip comes precoated with a substance that would react to an API (or an impurity) present in the drug. Upon reaction, the color of the paper strip would change, and the lack of color or discoloration of the paper strip would reflect a substandard product.[32] The advantages of the technology include ease of use, low costs, and ease of transportation. Technologies such as the one developed by researchers at Notre Dame, under the direction of Professor Marya Lieberman would fall into this category.

The third category of technologies consist of combination microfluidics and imaging systems, all condensed into a small portable and lightweight suitcase.[33] My own research group at Boston University has been involved in developing technologies that fit this category. Microfluidics as a technique has taken off within the last decade and deals with using very small volumes for chemical reaction and analysis, instead of liters or milliliters required in typical chemistry labs and assays.[34] The entire reaction can happen on a chip that is no bigger than a Band-Aid and allows the user to conserve resources and decrease the time of reaction. The idea in using microfluidic systems for drug-quality detection is based on the premise that specific molecules, which would be highly specific and sensitive to the active ingredient in the drug, can be developed. These molecules would react to the active ingredient in the drug, and upon the reaction it would give a light signal that is detectable using a smartphone or a simple camera. The amount of light that is given off is directly proportional to the amount of the active ingredient. Taking our example of drug Y earlier, this approach would rely on a molecule that reacts with the active ingredient in the drug Y. Upon reaction with drug Y, it would give off a light signal. The amount of light given off would tell the user exactly how much active ingredient is present in drug Y and how it corresponds to what the manufacturer said it should have. The advantage of these systems is that they are quick, quantitative, and work for various formulations, ranging from solid tablets, powders, and capsules to liquid injectables. The entire system can be fit into a small suitcase. The suitcase is fully self-sufficient and the technology requires minimal training. Additionally, it is fully automated and directly records the measurements in electronic format. The system is also capable of measuring drug dissolution, making it capable of addressing a major limitation of existing field-based systems. Nonetheless there are several limitations of this technology as well. The major limitation of the technology comes from the need to develop a large number of probe molecules for the various drugs that need testing in the field.

The fourth category of portable technologies on the horizon are based on the principle of high-quality imaging and recent improvements in cameras and image analysis. The US FDA has developed instruments using this approach. These technologies take a picture of the drug using cameras and lenses that are built into the instrument. The technology provides a high-resolution image of the drug that is able to capture even the minute physical features. By comparing the physical features of the drug to the authentic drug (an authentic reference for comparison is needed) the technology is able to discern whether it is real or not.[35] The user would use this kind of technology to compare drug Y that was bought from the market against the authentic drug Y provided by the manufacturer. If the drug Y from the market has suspicious spots, an unusual shape, a different color under a high-powered camera, or other physical features different from the drug provided by the manufacturer, it would raise a red flag. Thus in providing a high-resolution image, the technology is able to go beyond just visual inspection, or inspection of the box that contains the drugs. However, like other technologies it also has limitations. First, the technology depends on direct comparison with a standard provided by the particular drug company. In case of generics, this may not be entirely easy and would depend on generic manufacturers interested in collaborating. The authenticity is determined in a relative manner and hence the drug tablet from the manufacturer needs to be available. Second, the technology provides no information about substandard drugs that may be produced by genuine manufacturers. Third, it is unable to provide information about liquid samples as well.

In addition to these four broad categories of technologies, there are other efforts to decrease the costs of higher-end technologies to make them more accessible. There is an effort to decrease the cost of a mass spectrometer and reduce it to one tenth of the existing cost.[36] The mass spectrometer would still be fairly expensive (in the range of nearly a hundred thousand dollars) despite this massive reduction. Additionally, the need for quality infrastructure and the technical ability to maintain the instrument would still be needed.

All technologies, despite their promise, are inherently limited, and their ability to make an impact depends on the nature of the question asked. There is no one-size-fits-all approach that can be employed, though some sizes do fit better under certain circumstances than others.

In addition to the need of a technology and the needs of the user, there are also real challenges faced by those who are in the business of technology development. Academics and entrepreneurs look at the opportunities in emerging markets from different lenses. The developers, whether they are in academia or industry, face a trifecta of challenges.

The first challenge faced by the technology developers is that of awareness and context. Most literature available is on counterfeit and not on substandard drugs, which means that many technology developers are unaware of the complexity of the problem.[37] This is part of the bigger problem in global health and development, where until very recently, technology developers, engineers, scientists, and innovators have had a relatively smaller role to play.[38] The discipline of global health has historically been dominated by medical practitioners and public health workers, along with researchers from the social sciences such as economics, anthropology, and sociology. With the notable exception of environmental and civil engineering, in projects related to water and more recently the environment, the role of engineers in technology development for global health has been a relatively recent phenomenon. As a result of this historical gap, the level of awareness among engineers, technologists, and innovators is nascent. This gap of interaction and limited knowledge has resulted in limited understanding among technologists about the multidimensional and nuanced aspects of the problem.

The problem of awareness is not just present in the engineering and science programs in high-income countries. It is also acute in places where the problem is felt with the greatest pain. Students, entrepreneurs, and innovators in developing nations are largely unaware of the complexity of the problem and the innovations needed to address the challenges. Even in places such as India, Pakistan, and Ghana, where there is some recent activity in the innovation and entrepreneurial sphere, the approach is often highly dependent on capital and backing of investors from the developed world.[39]

The problem of awareness is itself multifaceted. Those who do know about the problem, despite little literature and lack of educational components, do so through media or ad hoc streams of knowledge. As a consequence, there is little understanding of the sociocultural context and the real challenges on the ground. Thus, those who can bridge the first gulf of knowledge and attempt to create solutions often do so in isolation and their solutions have limited application. The technology developed in a lab in Boston, for example, with little input from local partners on the ground, or lack of cultural understanding, is more likely to fail than to succeed.

A poor understanding of local context among technologists and engineers is a big problem, and not just present in developing technologies for substandard and counterfeit drugs. It is a problem that plagues developing technologies for global health settings in general.[40] Because some of the major challenges in global health, such as malaria, HIV, TB, and so on, have gotten more attention, both in the academic (scientific) and popular press, understanding of the technological needs associated with these diseases is more sophisticated than the challenges of poor-quality medicines, where the prevailing opinions surround only counterfeit drugs, with little understanding of the complexity of substandard drugs.

Recent efforts, driven in part by international consortia, have aimed to improve the overall awareness of the problem—however, how much of that has penetrated technology development remains unclear. A key driver for improving awareness among engineering researchers and entrepreneurs is through research grants and funding. That pool of money is very small to begin with and does not seem to be growing at the rate it needs to grow.[41] There are very few, if any, grants available to encourage researchers to address the problem. In the absence of grants, the initiative by researchers to address the problem remains weak.

Those researchers and technology developers who are somehow able to overcome the first two barriers of awareness and funding to do early stage pilot development face an even bigger challenge of scaling. The typical financing models available for the scale-up of innovations are developed for markets in developed nations,[42] where the cost recovery and returns on investment models are better understood, taught in business schools, and practiced widely. When it comes to markets in developing nations, despite the population size and the potential market, investment in the scale-up of innovation is full of risks. These risks stem from political turmoil to corruption, from breach of intellectual property to physical security of the institution. The inherent insecurity of investors is further compounded by a lack of successful models for scale-up. The investment needed for a technology that requires hardware (which is distinct from an app or a smartphone platform) is substantially higher, and it remains unclear who will be the final payer and how the costs will be recovered or sustainability will be ensured. The poor interactions of global health technologists with the business and investment community, and the wide gulf that exists between them, also negatively impact the development of technology.

Further complications are created by lack of clear policy, or any regulation whatsoever, in the area of medical devices and technologies in developing nations.[43] This lack of regulation is distinct from regulating medicines' quality and is concerned with the laws associated with medical device import, performance, licensing and sales, and protection of intellectual property.

Regarding substandard medicines, the role of the local government cannot be understated. Any technology that is aiming to address the public health challenges created due to poor-quality medicines will need government support for testing, scaling, and incorporation into existing systems. With tremendous variation in medical device laws (and, in many cases, the complete absence of any policy on medical devices whatsoever) the lack of government will, action, or understanding of technology makes the problem further complicated. Often times the law that regulates medical devices and their classification is poorly written and even more poorly understood.[44]

The issues of intellectual property from the technology perspective are also substantial. While a lot has been written about the various treaties around generic drugs and the role of the World Trade Organization and its implications on countries like India, the intellectual property associated with technology to test drug quality is poorly understood.[45] From the developer's perspective, intellectual property is highly valuable in ensuring the ability of the developers to raise funds, get investor backing, and have the means and mechanisms to continue to support innovation. Yet developers and their institutions may be reluctant to pour in the resources for patents that may not be fully protected in low-income countries and may have a high chance of infringement.

The long-term sustainability of the technology also depends on the financial model of the enterprise that would develop the technology and support it in the field. Here, it is not only the cost of technology infrastructure, maintenance of a robust R&D culture that continues to create new solutions to new challenges while minimizing the cost per test, but also the question of who can, or who should, pay for testing? At the core of this issue is the fundamental question of user versus payer.

In the case of smartphone-based models, the financial model is to have either large pharmaceutical companies concerned about their brand, or mobile phone companies interested in corporate social responsibility, pick up the cost. The user here is the person who is going to consume the medicine and make the call with the toll-free number. The cost to the user is not in terms of cash, but in terms of time and effort to make the call. It may also come in terms of time and effort that is needed to return the medicine, should it prove to be a counterfeit. In this model, either the government has to provide subsidies and pay the private companies to put the barcode on the drug, or the drug companies have to see a clear benefit to continue to support this program. There is also good publicity and aspects of corporate social responsibility for mobile phone companies.

Beyond smartphone-based solutions, the business model gets more complicated for other technological solutions. For large-scale testing of various brands, it is unlikely that the manufacturer will pay for all tests. This is also true in the case of generics, where generic firms may not have an interest to do large-scale, countrywide testing of their drugs. Expecting the consumer to pay the costs is not viable given the limited financial ability of the consumer in these environments. In the absence of universal healthcare models and a lack of insurance mechanisms, the consumer is also unlikely to recover his or her costs should the drug turn out to be of poor quality.

Other financial models may include the idea used by razor-blade companies and some printer and cartridge companies.[46] The idea is to keep the cost of the printer (or the razor) minimal but generate revenue from consumable parts (e.g., the replacement blade or printer cartridge). A similar model may be created where the cost of the hardware could be low (or nonexistent) and the user will pay for the consumables. The user in these models is assumed to be the government, the regulators, or private entities that have the ability to pay on a regular basis. These kinds of strategies have been used with imaging-based systems where the reference libraries have to be purchased individually.[6]

The new technologies, coming up on the horizon, work on the principle of low costs per test and negligible cost for hardware and infrastructure. The assumption is also that not the patient (or the buyer of the drug) but some other entity, interested in maintaining the overall quality of the medicines in the country, will be the user and the payer of the technology. Because the cost per test is low by design, the financial success of these technologies will therefore depend on the high number of tests carried out. This means several things have to change from the current situation. One possibility could be that the government would conduct regular and frequent testing and in substantially higher numbers than it does right now. This would allow for cost recovery for the firm that has invested in creating the technology. However, additional tests would also require hiring of new staff by the government, or the tests would have to be carried out by the private-sector health facilities. Both of these requirements would imply investment of resources from the government, which in a developing country is inherently challenging. The government in a low-income country has limited resources and a higher number of tests would require substantially large investment. Thus the technology companies will have to then make up the numbers by operating in a large number of countries, which, given the country-to-country variation, is a challenging proposition.

The alternative to public-sector investment is testing in the private sector, such as private hospitals or private pharmacies. This model, while having some promise, is yet to be fully explored. In a number of countries, the private sector does not have the license to buy the medicines directly from the manufacturer and instead uses the government warehouses and facilities to procure drugs. Thus the private hospitals may put pressure on the government to test and ensure quality, rather than having to test the drugs themselves. Because they may not be able to procure drugs directly from the manufacturer, who is likely to be outside the country, finding that the government provided drugs are of poor quality would not help them and are likely to create a stock-out. In these cases, the investment in drug testing for private hospitals would not make a lot of financial sense. Second, in the case that the drug is found to be compromised, the private sector would need to have some authority or channel to put pressure to recover their costs to get part of the fines. This would mean creating a mechanism by which the government would reimburse them in case the drug is not of the desired quality.

The case with private pharmacies and chains of pharmaceutical stores is slightly different and potentially promising. Yet their ability to test would need to rest on the promise that they can test before they pay for the shipment, and it would have to be agreed upon by their suppliers. The suppliers will have to have some kind of assurance from their providers that their costs would be recovered in case the drugs are not of the desired quality. However, there is an upshot to this as well. If the pharmacists and even the hospitals are able to argue and advertise that their drugs are not just from good brands, but also have been tested through a reliable technology, it is likely to increase their trust and hence their customer base.

Despite the myriad of challenges associated with technology development, implementation, and scale-up, over the last few years, there has been an increase in research and entrepreneurial activity associated with development of new technologies. The main reasons for this positive (albeit small) advancement are slightly increasing awareness, some outreach campaigns, and a few new funding opportunities.

With regards to awareness, despite the relatively poor communication between public health institutions and engineering programs, efforts by the Gates Foundation, USAID, and other partners in Norway, Canada, the United Kingdom, and Korea have emphasized the need for innovation, particularly around maternal, newborn, and children's health.[47] The problem of substandard drugs is, of course, much bigger than just newborn and maternal health, but the attention from aid agencies to innovation is increasing interactions among engineers, entrepreneurs, and public health professionals.

Perhaps a bigger impact is coming through the creation of new funding streams. While a new announcement for long-term commitment to funding, specifically focused on developing technologies for substandard and counterfeit drugs, by any major federal agency in the United States or elsewhere, has not come through, various funding agencies have broadly defined their public health request for applications, hence making it possible for innovators to consider applying for these programs. In this respect, the Saving Lives at Birth Consortium and the United States Pharmacopeia (USP) Fellows program[48] are particularly noteworthy. These programs offer student stipends (USP fellowship) and research funding (Saving Lives at Birth) to improve the health system and drug quality in low-income countries through innovation. While a step in the right direction, these efforts need to be multiplied several fold and in several countries to foster creativity and sustain innovation.

Some private investors as well as the nonprofit sector have shown an appetite for mobile platforms, including the ones developed by mPedigree and Sproxil, however, the appetite among these investors to tackle the problem of substandard drugs and comprehensively improve in-country pharmacopeia remains limited.[49]

An analysis of the technology development landscape, from the perspective of innovators, paints a complicated, and at times grim, picture. There are systematic barriers to technology development, testing, and scaling, and not all of these barriers are associated with the technology itself. Many are due to the lack of financial models for global health technologies and some are there due to lack of awareness.

Overall, the lack of appropriate technologies that can address the problem is reflective of the complex landscape of drug quality in which these technologies need to strive and thrive.

10

The Ivory Trade

It was March 2014, and my research team and I were in Accra to do field tests of the technology developed in our lab to test the quality of malaria medicines. It was my first time in Ghana. Our mission was to understand the local health system and to test our technology outside of our lab in Boston, with local staff and locally procured drugs.

On the first night in Accra, we had met a fairly knowledgeable and pleasant taxi driver named Ahmed. Originally from Burkina Faso, he had been living in Ghana for nearly two decades. As we went to a popular Turkish restaurant, upon the recommendations of some friends and colleagues, Ahmed and I hit it off immediately. The next morning, there was a small window of time that I had available, to see the city, before my team and I were going to start testing our technology. I decided to see what the urban slums of Accra look like.

I called up Ahmed in the morning and asked him to take me to a slum. There was some degree of confusion and hesitation on his part. First, he did not know what I meant by a slum. After I tried to explain to him what I meant, he had a much bigger question. Why would I want to go there? My reasoning, that I wanted to see the local pharmacies there, was not particularly convincing to him. If I wanted to go to a pharmacy, he told me, I should go to the one close to the mall: that was a better choice.

They had much better and safer drugs, and I should avoid getting drugs from random places, he said trying to protect me. Our conversation went back and forth, and eventually he reluctantly agreed. I am convinced that he had serious questions about my judgment.

Driving from our hotel, which was close to the airport, it took us nearly a half hour to get to the area called Agbogbloshie and the informal settlements around it. One of the world's biggest dumping grounds for electronic waste (e-waste) and hardware, called Gomorrah, is in this region.[1] I did not have the time to go in the e-waste dumping grounds, but I did have time to walk around and see the formal and informal pharmacies and the bustling trade that was going on. In dusty old shops, there were people coming and going. There was no electricity on the day I went there, yet the attitude was that of a bustling market. Plenty of people were buying and selling, some in shops, some from carts, and some from vendors sitting on the street. There were shacks that were selling pretty much everything that one could imagine. Also in that environment of active business were sewage pipes that were open, plenty of stagnant water, dust, fumes from the cars, and food being sold on the street. The air was thick with these mixed scents and smells. I went to several pharmacies and inquired about the various kinds of antimalarials that were available. My aim was not a sting operation, but just to find out what was being sold and for how much. Not everyone spoke English, but those who did showed me a variety of antimalarials that they were selling in their shops. There were significant price differences as well. Shops that were a stone's throw from each other and were selling the same brands, sometimes differed by as much as 20 percent in price. Some batches looked new while others were in boxes that were thick with dust. None of the shops had any real windows or ventilation, and with no electricity it was mostly dark inside.

My colleagues, on the other hand, had arrived at CePAT. CePAT stands for the Center for Pharmaceutical Advancement and Training. Housed in a facility about five kilometers from the picturesque campus of the University of Ghana, CePAT, started in 2013, is in a league of its own in terms of its infrastructure and resources. Built through the help and vision of the US Pharmacopeia (USP), it is a first international venture of its kind. The USP CEO, Ron Piervincenzi, recently described the center as "a platform to train African professionals and build capacity for domestic and sustainable drug quality systems. Since 2013, CEPAT has helped train 190 professionals from 32 African countries."[2]

CePAT houses state-of-the-art equipment and provides technical support, training, and testing services of high international standards. More than just a training center, CePAT has also started to develop partnerships, across Africa, to engage the local governments and to encourage them to improve their capacity. The day my team and I were doing testing, regulators and inspectors from half a dozen African countries, from Kenya to Mauritius were getting trained. The training manuals were detailed, the exercises rigorous, and everything was running strictly on schedule.

CePAT was ten kilometers from the settlements of Gomorrah, but it was a world away.

The presence of CePAT and the informal markets peppered throughout Accra tell a story about the challenges and the promises that coexist side by side. It is a reminder of how deeply rooted the challenge is, and why there is reason to be cautiously optimistic.

Perhaps the most important development is the broadening of the discourse from counterfeits to substandard drugs.[3] The expansion of the vision, from not just catching those who break the patent law and trademark agreements, but also protecting the vulnerable who suffer because of negligence and complacency, is a major step.

Recent conversations both within countries and on the international stage no longer look at counterfeit drugs but now include the real impact of substandard drugs. Researchers in the regulatory sphere, academia, and the public health community have started to clearly articulate that substandard drugs deserve equal attention to counterfeits, as they pose a great, if not a greater, threat to public health.[4] Recent publications, journal articles, conferences, and public awareness campaigns have started to make this distinction. This intellectual departure has been critical to motivate governments to invest in improving health and regulatory systems, increase public awareness, and also create incentives for new technologies that can address problems at the point of care and the point of need. Research-based pharmaceutical companies, which have historically invested heavily in anticounterfeiting measures that had to do more with brand protection, have also shown a strong willingness and interest to increase awareness about poor-quality products that can have disastrous consequences on human health. News publications and magazines in recent years have also embraced the nuanced understanding, and have used clearer language to illustrate the various categories of bad drugs, along with their impact on health systems and disease-control programs.

There have been international efforts, combining expertise in public health, pharmacy and regulatory sciences. Among the programs aimed at addressing the issue of capacity building, awareness about substandard drugs, and partnership with the Ministries of Health is the Promoting Quality of Medicine (PQM) program from the United States Agency for International Development (USAID).[5] The program is based on a partnership between the USP and USAID, particularly with the office of health system strengthening at USAID. The current PQM program builds on nearly two decades of partnership, collaboration, and in-country support.

USP, over the last quarter century, has increased its international efforts aimed at improving the quality of medicines through awareness, advocacy and capacity building. In addition to its own efforts in global health, which have grown considerably over the last decade, its partnership with USAID has substantially contributed to the awareness of quality issues.[6] The USP-USAID partnership in countries where USAID works has taken a systems-level perspective with a focus not just on awareness about quality issues but also about medicine procurement, drug registration, and resource allocation.

The USP-USAID partnership, and a joint commitment to tackle the problem of poor quality drugs, laid the foundations for a program called the Rational Pharmaceutical Management (RPM) project.[7] The goal of this project was to develop easy-to-access yet comprehensive and accurate monographs about drugs that were part of the World Health Organization (WHO) essential medicines list. A major success of the RPM project came in Nepal, where information centers with accessible information and monographs on drugs and drug quality became a major resource for institutes and units within the Ministry of Health, and in doing so a culture of collaboration, information, and resource sharing arose.

The success of RPM helped launch the Drug Quality and Information (DQI) program, which was created in 2000, through continued partnership between USP and USAID. With HIV/AIDS being a major global concern at the time, along with malaria and TB, the DQI program also expanded its scope and vision.[8] The goal of DQI in Africa, Latin America, and Asia was not only to simply provide monographs, but also to start assessing the existing quality control measures in these regions. This invariably meant that the scope would include discussion about counterfeit and substandard drugs, as well as monitoring and evaluation of the current programs to protect public health that is vulnerable due to the continued challenges posed by poor-quality medicines. The DQI program created several programs, including seventeen in Latin America in seven countries that were focused on antimalarial quality.[9] At the same time, this program started technical training of the local staff and emphasized the importance of field testing for efficient quality control. The DQI program provided additional support to improve drug-registration systems to protect against fraud. The DQI support helped establish Madagascar's National Center for Pharmacovigilance,[10] which was subsequently accepted as a member of the WHO International Drug Monitoring Programme. Similar efforts in the Greater Mekong Subregion, focused on antimalarials, resulted in new partnerships and increased capacity.[11] The increase in surveillance and awareness has led to the expansion of the program to antibiotics, anti-TB medicine, and antiretrovirals and have subsequently led to a partnership with Interpol and successful raids that seized US$9 million of counterfeit drugs.[12]

Based on the success of the DQI program, a new comprehensive PQM program was launched in 2009. Started with US$35 million, the program funding has now increased to $110 million and the program will stay in operation until 2019.[13] The PQM program is also unique as it works independently from the US government and is based on technical expertise from USP. USAID in creating this new venture with USP stated that addressing the issue of poor-quality medicines is central to its global health mission, and addressing it comprehensively, through partnership with

USP, is a cornerstone of its mission. It argued that, "without it, Ministries of Health, regulatory authorities, manufacturers and other international stakeholders will have no economic or political interest in working with USAID in this field."[5] The PQM program, through this partnership and expanded vision, works in thirty-five countries around the world.

The PQM approach is unique in several ways and presents a marked departure from other international and collaborative efforts. First, it is focused not just on counterfeits but also on substandard drugs. This focus has enabled PQM to work with local, national, and regional authorities to emphasize the issue, improve training in quality assurance and quality control, and take a systems-level perspective. This has also led to in-country capacity building, improvement in surveillance, and registration of foreign-manufactured drugs to ensure quality, access, and availability. This is connected to the efforts associated with working with the public sector and local FDAs to enact better policies to protect local pharmacopeia.

The second important tenet of the PQM program is emphasizing good manufacturing practices with local manufacturers. This is important in not only strengthening the local manufacturing base, but also improving the overall quality for in-country supplies. This also improves the partnership between various local pharmaceutical manufacturing, supplies, and access sectors.

A third aspect is PQM's training of in-country staff. The training, conducted in local languages and adapted based on the skill level of the participants, is a major part of PQM activities. These training activities focus not just on protocols, but also on equipment usage and technical understanding of the analytical processes. This ensures that the staff being trained not only become aware of key aspects of counterfeit and substandard drugs, but also recognize the technical aspects of detecting them. To ensure that the training program benefits the entire system, PQM also focuses on sampling methods and emphasizes how various components of the system, from sampling to field testing to lab testing, are supposed to work in synergy. Additionally, sustainability of the program is a major goal and is achieved through a train-the-trainers approach.

While training exercises and refresher courses aim to build local capacity, they are part of the bigger effort to improve in-country drug testing and medicine quality control labs. These labs, which are a central part of a country's effort to test drugs, provide evidence and protect public health and are often in need of technical and logistical support. PQM aims to fill these gaps through technical training of staff and awareness about current technologies. At the same time, PQM also provides recommendations for technologies that the labs can purchase for specific needs.

Fourth, the PQM program offers a Visiting Scientists Program to bring technical staff from partner countries to USP headquarters in the United States and to introduce them to various technologies and protocols for training on good laboratory practice, quality control and quality assurance, testing, and analysis. The visiting scientists are then expected to go back to their home countries and train their colleagues.

Another key aspect of PQM activities, with the support of in-country USAID missions and the local ministries of health, is to analyze the existing systems in the country and to identify gaps in regulatory practices, surveillance, capacity, monitoring, or technical expertise. PQM then makes recommendations to the appropriate authorities on how to fill those major gaps in knowledge, capacity, or practice, keeping in mind the local challenges, resources, governance structure, and existing capacity.

The process of quality control is long and multifaceted. Support is not only needed to initiate a program and develop initial capacity, but continuous support at each step is equally vital. As a result, PQM has to ensure that there is concerted effort to provide mentorship, advice, and technical support once the in-country quality control programs are established. PQM has partnered with WHO to improve pharmacovigilance using the Medicines Quality Database (MQDB) program. The goal is to provide access to users to search the MQDB database by location, time period, type of medicine, and quality test results and understand the local, regional, and global challenges that may further improve their own practices and policies.[14]

USAID and USP are also increasing their efforts to identify gaps in existing technologies.[15] Partnerships with universities, federal agencies, and innovators are being developed to fill gaps in the technological space.[16] While the effort is at an early stage, and technology development takes a long time, the mechanism is increasing interactions between various stakeholders in technology development and testing, and increasing awareness of the challenges among innovators, engineers, and technology developers. At the same time, it is also bringing new awareness among local regulators about the process of technology development and ensuring that they have a say in what the in-country needs are.

The final component of PQM's activity is raising awareness among not just those whom it trains, but also other important stakeholders in public health, including pharmacists, local healthcare providers, physicians, nurses, and the community in general.[17] These efforts, driven through traditional media outlets and more recently through electronic and social media, are aimed at improving general awareness and promoting good practices in procuring prescription drugs from licensed pharmacies.

The issue of awareness has also been taken up by the International Federation of Pharmaceutical Manufacturers Association (IFPMA) and its partners through its Fight the Fakes campaign. Started in late 2013, the program is a recognition that awareness and engagement remain the ultimate challenge and also the best opportunity to create global best practices.[18] The goal of this campaign is threefold. First, it is to increase awareness among various groups that have a stake in improving the quality of medicines, which includes not only the manufacturers, pharmacists, suppliers, and physicians, but also the general public. The second component of the campaign is to create a global alliance, through partnerships between organizations, institutions, and civil society groups to improve awareness and encourage best practices that safeguard the society from the challenges of poor-quality medicines. Third, the Fight the Fakes campaign aims to become a global hub of reliable information, and a platform of information sharing, that is reliable, robust, and easy to access for various groups based on their knowledge, skill level, interests, training, and vision.[19]

As of early 2016 the campaign had thirty-two partners around the world.[20] A major component of this campaign is its online presence. The website of the campaign aims to engage users with various backgrounds and interests, and it regularly posts stories of individuals at the frontline, or those who have suffered from adverse reactions of poor-quality medicines to increase awareness. Similarly, it provides access to recent news about developments in the field, as well as changes that positively or negatively impact the global fight against poor-quality medicines. Fight the Fakes, while initially housed in the IFPMA (and moved to International Federation of Pharmaceutical Wholesalers, or IFPW in 2017[21]), tries to have a much bigger footprint than the hosting organization. Their posters and booths are present not just at IFPMA meetings and symposia but also at other meetings of the UN, the WHO, and other relevant conferences on tropical medicine. Recent efforts have also recognized the need to engage students and there is an active effort to have local chapters at universities to improve awareness among students. The partnership with University College London is a step in this direction.[22] Yet a lot more needs to be done, at not only institutions in Europe and the United States (where the presence at university campuses is still minimal), but also at local campuses in low- and middle-income countries. Given the short history of the effort and the changing dynamics of media, it is hard to exactly ascertain the impact of the campaign on the ground. While the early signs on social media do indicate a positive trend, in terms of increasing awareness, the effects of this and other similar campaigns can only be felt in the long term. It is therefore important to provide continued support, space, and resources to these campaigns for them to be able to create sustainable impact.

Fight the Fakes represents one of the several efforts in the contemporary media world. There are other bloggers who have taken up the issue as well to increase awareness and disseminate information.[23] These bloggers are driven to the cause because of personal experiences or passion, and their effort varies depending on personal interests, focus, and vision. While there are many who are focused on the issue of counterfeits, since that continues to be a major component of the discussion, there are some recent efforts that are also highlighting substandard drug challenges.

It is important to note that the effect of the Fight the Fakes campaign, or that of other blogs, is not to build new technologies or to provide training to the ministries, but instead to engage the generation that may look to blogs, Twitter, and Facebook for their information. Their presence, if done with regard to accuracy, can improve general awareness. That said, there is always a danger about the dissemination of news that is inaccurate or outright dangerous. Engagement of these various electronic and social media entities with reliable sources (such as the US or European Pharmacopeia or reliable Medicine regulatory authorities) can further strengthen the message and create new channels of collaboration that are mutually beneficial to large organizations and bloggers.

The technological sector has also started to see more activity, driven in part by awareness among the innovators and in part by the presence of new grant mechanisms to foster innovation. There is growth in the academic sector as well. Major international conferences on global public health that previously focused almost exclusively on the public health angle or policy efforts to tackle the issue of substandard drugs now have sessions on technological innovation needed for improving drug quality.[24] Further growth in these sessions at conferences can bridge the intellectual gulf between engineers and technologists and those who work at the frontlines. This growth will also allow for rigorous discussion on where the technological gaps are and what to do about it. Similar scenarios are reflective in research publications of journals that also occasionally feature technology innovation.

The creation of CePAT, expansion of PQM, efforts by Interpol and law enforcement authorities, and awareness campaigns all point to steps in the positive direction. But these are few compared to the size and enormity of the global challenge: a lot more needs to be done. More international programs like PQM, more labs like CePAT, and more local ownership of innovation is needed. Similarly, coordination between various arms of awareness, collaboration, and data sharing is also required for these efforts to be internalized and become sustainable.

Stronger international partnerships are also echoed in the sentiments of legal scholars and economists who study the problem. Amir Attaran, a legal scholar at the University of Ottawa, who has studied substandard and counterfeit drugs for quite some time, argues for a strong partnership beyond the few countries that we see on the global stage. A global framework or a legal treaty, he argues, is probably the way to address the issue. In a recent interview to *Newsweek*, he proposed creating a treaty that would be followed by all. He gave the example of civil aviation as an analogy: "There are dozens of treaties on civil aviation, and every single country is following those. If not, they don't fly."[25]

Roger Bate, another noted scholar who has written extensively on this topic, argues for stronger medicine regulatory authorities. Because medicine regulatory authorities are the link between both substandard and fake medicines, they are positioned to make a long-term and sustainable impact through enforcement of stricter standards. He argues that even in poor countries, it does not require much work to ensure that laws against counterfeiting are in place.[26]

In the end, it boils down to awareness that corresponds with a deep commitment to action. The issue of awareness is more than just telling the public or the clinicians the harms of poor-quality or fake drugs. There are awareness challenges even among those who have been working in the field for decades. For example, awareness about both the technological limitations and the need to create new technologies is desperately needed among regulators and public health professionals. Technology development requires both upfront investment and significant resources to shepherd it through various stages of testing, optimization, and implementation. That process cannot be undertaken purely through donations or development funds from private organizations. Technology development requires long-term funding and the public investment is necessary. The technology investors in the developed world, even those who care about health and well-being, lack awareness about global health challenges. In general, there is a serious gap in funding mechanisms for new technologies. While there is a significant amount of money that is being spent on new initiatives for capacity building and for local training, the emphasis on creating new technological solutions, or optimizing existing ones to address the critical needs, is lacking. Existing researchers and entrepreneurs in this area struggle to keep their efforts going and the incentives for new scholars, innovators, and scientists are minimal. A global consortium, such as a global fund, or a similar initiative can dramatically change the landscape of technology and accessible solutions through investment in this arena. The private sector does not offer much either. Investors use their lenses of the developed world to evaluate the success of technologies in the developing nations. While the returns on investment in health in general can be promising, the short term or immediate returns are unlikely. Those who are in charge of funding in the public sector need to be made aware that technology needs for the developing world are different, and so are the risks and opportunities. Policy experts need to be made aware of the local innovation culture and the gaps that need to be filled in that area.

But innovation is not just limited to technology alone. Better financing models for existing solutions and an understanding of the scale-up of efforts, whether they are technological or social, are equally needed.[27] Here, academia needs to be both proactive in creatively looking for solutions, and making new platforms accessible to the students and researchers who wish to contribute.

Awareness, at all levels, in all dimensions, remains the most potent tool.

In March 2013, Chinese President Xi Jinping went to Tanzania. The visit was focused on bilateral ties as well as Chinese investments and contribution to the development of Tanzania and other nations in the region. As state visits go, after the fanfare the headlines went into the background. Except that this time, they resurfaced in November 2014. Reports started to emerge that some of the high-ranking officials traveling with the president bought illegal ivory from Tanzania and smuggled it by putting it in diplomatic bags that do not go through customs and regular scrutiny.[28] The news made headlines, appearing in the press from Africa to America, and from Australia to India. This was a major embarrassment for China, which in the last few years had tried to demonstrate its resolve against poaching and illegal trade in ivory and rhino horns. China has taken a strong official stance to position itself as a serious international player in the conservation movement. The situation was particularly sensitive because Tanzania had been the worst hit by the poaching crisis and estimates suggest that it could lose all of its elephant population within the next decade.[29]

China strongly rejected the accusations and called them baseless and highly irresponsible,[30] but the damage was done. There was an outcry and major international embarrassment.

The issues of poaching and ivory resonate with people from various social sectors and in a vast number of countries. It has gone from an issue that is unique to Africa and perhaps China to conversations about conservation. *National Geographic* and other environmental groups have made this an important topic worthy of their effort and outreach, even to young kids.

Ivory trade and poaching are also considered morally and socially unacceptable, and there are international ads and awareness campaigns about them. For example, the campaign against ivory appears on various ads and signs at various airports around the world, including at airports in China. The issues is often featured in ads in popular magazines featuring sports and movie celebrities taking a strong stance against the ivory trade.

I was in Tanzania in early January of 2015, running from one meeting to another at the drug and poisons board and visiting the office of the anticounterfeiting agency. The story about China and ivory was still a topic of discussion in many circles. Talking about drug counterfeiting, against the backdrop of ivory trade and global shaming, I realized that there is an important lesson to be learned. The drive for greed, the complacency and corruption at various levels, and a lack of good surveillance are all similar challenges. Just as a global campaign needs a strong legal framework and an agreement to protect the lives of the most vulnerable, the campaign against drugs that deliver the exact opposite of health and wellness needs effective communication that speaks not just to those who have suffered but also to those who care about the lives of the vulnerable.

The problem of poor quality drugs is truly international, and so should be the message to protect the vulnerable. A campaign that includes awareness and reaches people of different backgrounds according to their knowledge, interest, and education, and encourages them to take action, no matter how small, can create sustainable change. Campaigns that enable anyone and everyone to act, be it a child, a scientist, a stay at home parent, or a politician, can shape the international discourse and make knowledge accessible. This would lead to tangible international pressure. As the international public pressure will grow, so will the positive changes in policy and implementation practices.

On December 31, 2016, China announced that it will shut down all of its ivory trade by the end of 2017.

Epilogue

Thirty years later, D. Watson's is now a much bigger brand than it ever was. It has expanded to stores across the city. The small shop near the supermarket now occupies multiple stories, and even bigger stores are cropping up elsewhere around town, including one that is much closer to the home I grew up in. It is no longer a store that just sells medicines, though pharmaceutical sales are still part of the business. D. Watson is now a superstore, selling groceries, pet-grooming products, and even electronics in neatly stacked aisles. They are further expanding their market and will be carrying name-brand shoes and boutique clothes.

Their selling point, as it was in the 1980s, remains the same: Trust.

References

Chapter 1

1. *Ghanian Times*. Fire ravages Tema Medical Store. 2015-01-14 (2015).
2. KEMSA. *Historical Perspective on Kenya Medical Supplies Authority*. (2016). http://www.kemsa.co.ke/about-us/historical-background/
3. Personal communication with Gerard Requin, Former Director, Pharmaceutical Services, Mauritius.
4. Gamal Khalafalla Mohamed Ali, NMSF, S. D. G. National Medical Supplies Fund Sudan: Historical Background (2016).
5. Nest, G. H. Central Medical Stores fire, how it all happened. *Ghana Health Nest,* Jan 29 (2016). http://ghanahealthnest.com/central-medical-stores-fire-how-it-all-happened-govt-statement/
6. Saleem, A. U. S. PIC free medicine: As deaths soar past 80, authorities still clueless. *Express Tribune,* Jan 25 (2012). https://tribune.com.pk/story/326824/pic-free-medicine-as-deaths-soar-past-80-authorities-still-clueless/
7. BBC. Lahore grieves over heart pill deaths. 2012-01-29.
8. Chaudhry, A. Tests in UK identify drug which caused havoc in Lahore. *Dawn*. 2012-02-02.
9. Chaudhry, A. Drugs reaction: Samples of blood sent abroad amid blame game. *Dawn,* Jan 31 (2012).
10. The Pathology of Negligence. Report of the Judicial Inquiry Tribunal to Determine the Causes of Deaths of Patients of the Punjab Institute of Cardiology, Lahore in 2011–2012. (2012).

11. *DRAP Act, Government of Pakistan.* (2012).
12. *New York Times.* Cough syrup suspected to have killed 33 in Pakistan. 2012-12-29.
13. Jaffrey, S. Deadly risks run by Pakistan's cough syrup addicts. *BBC News.* 2013-02-07.
14. Chaudhry, A. Blame game? *Dawn.* 2013-01-20.
15. US Dept of Justice. *Cape Cod man indicted for trafficking in counterfeit Viagra.* 2014.
16. Faucon, B., Murphy, C. & Whalen, J. Africa's malaria battle: Fake drug pipeline undercuts progress. *Wall Street Journal.* 2013-05-29.
17. Bad medicine. *Economist.* 2012-10-13.
18. Committee on Understanding the Global Public Health Implications of Substandard, Falsified, and C. M. P. B. on G. H. I. of M. & 20., W. (DC): in *Countering the Problem of Falsified and Substandard Drugs* (ed. Buckley GJ, Gostin LO, E.) (National Academies Press (US), 2013).
19. Voice of America News. Malawi public hospitals face acute drug shortage. 2013-02-14.
20. Govindaraj, R. & Herbst, C. H. *Applying market mechanisms to central medical stores: Experiences from Burkina Faso, Cameroon and Senegal.* World Bank, (2010).
21. Ghana Ministry of Health. *Central medical store to be rebuilt.* Report from Pharmaceutical Society of Ghana, 2016.
22. PIC scandal to bring more problems for government. *Pakistan Today* 2012-02-12.
23. Supreme Court disposes of case against Efroze Chemical. *Pakistan Business Recorder* 2014-03-20.
24. Efroze Chemicals website. Available at: http://www.efroze.com/.
25. Hoodbhoy, P. Promoting anti-science via textbooks. *Dawn* 2016-12-03.
26. Farooq, U. Hepatitis C treatment: K-P hospitals receive interferon after two years. *Express Tribune* 2015-03-29.
27. Commission of the US FDA. *Counterfeit Drugs: Fighting Illegal Supply Chains* (Office of the Commissioner, 2014).
28. Wanjau, K. & Muthiani, M. Factors influencing the influx of counterfeit medicines in Kenya: A survey of pharmaceutical importing small and medium enterprises within Nairobi. *Int. J. Bus. Public Manag.* **2**(2), 23–29 (2012).
29. Health sector after 18th Amendment. *Dawn* 2011-07-10.

30. Perry, L. & Malkin, R. Effectiveness of medical equipment donations to improve health systems: How much medical equipment is broken in the developing world? *Med. Biol. Eng. Comput.* **49**, (2011).
31. Personal communication, Drug Regulatory Authority of Pakistan. (2015).
32. Chukwudi, N. H. Future dimensions in pharmacy practice for developing countries: A Nigerian perspective. *Nov. Sci. Int. J. Pharm. Sci.* **1**, (2012).

Chapter 2

1. Fernandez, F. M. et al. Poor quality drugs: Grand challenges in high throughput detection, countrywide sampling, and forensics in developing countries. *Analyst* **136**, 3073–3082 (2011).
2. Newton, P. N., Green, M. D., Fernández, F. M., Day, N. P. & White, N. J. Counterfeit anti-infective drugs. *Lancet Infectious Diseases* **6**, 602–613 (2006).
3. Wanjau, K. & Muthiani, M. Factors influencing the influx of counterfeit medicines in Kenya: A survey of pharmaceutical importing small and medium enterprises within Nairobi. *Int. J. Bus. Public Manag.* **2**, (2012).
4. Corfield, P. J. From poison peddlers to civic worthies: The reputation of the apothecaries in Georgian England. *Soc. Hist. Med.* **22**, 1–21 (2008).
5. Gevitz, N. In *Apothecaries and the Drug Trade: Essays in Celebration of the Work of David . . .* (eds. Gregory J. Higby & Elaine C. Stroud) (American Institute of the History of Pharmacy, 2001).
6. Barrett, C. R. *The history of the Society of apothecaries of London* (E. Stock, 1905).
7. American Academy for Advancement of Science. *Historical Trends in Federal R&D | AAAS—The World's Largest General Scientific Society* (2017).
8. Young, D. C. *Computational Drug Design: A Guide for Computational and Medicinal Chemists* (Wiley, 2009).
9. Seoane, J. A. et al. Biomedical data integration in computational drug design and bioinformatics. *Curr. Comput. Aided. Drug Des.* **9**, 108–117 (2013).
10. Mullard, A. Parsing clinical success rates. *Nat. Rev. Drug Discov.* **15**, 447 (2016).

11. Socolar, D. & Sager, A. Pharmaceutical marketing and research spending: The evidence does not support PhRMA's claims. *Am. Public Heal. Assoc. Annu. Meet. Atlanta* (2001).
12. LaMattina, J. Universities stepping up efforts to discover drugs. *Forbes* 2013-10-21.
13. Kaplan, W. A. et al. The market dynamics of generic medicines in the private sector of 19 low and middle income countries between 2001 and 2011: A descriptive time series analysis. *PLoS One* **8,** e74399 (2013).
14. Deshpande, R. Cipla. *HBS Case Study* (2003).
15. Chopra, S. S. et al. Industry funding of clinical trials: Benefit or bias? *JAMA J. Am. Med. Assoc.* **290,** 113–114 (2003).
16. Sheiner, L. B. & Rubin, D. B. Intention-to-treat analysis and the goals of clinical trials. *Clin. Pharmacol. Ther.* **57,** 6–15 (1995).
17. Piachaud, B. S. Outsourcing in the pharmaceutical manufacturing process: An examination of the CRO experience. *Technovation* **22,** 81–90 (2002).
18. Securing the pharmaceutical supply chain Hearing of the committee on health, education, labor and pensions. US Senate. One hundred Twelfth congress, First session. 2011-09-14.
19. Yu, L. X. Pharmaceutical quality by design: Product and process development, understanding, and control. *Pharm. Res.* **25,** 781–791 (2008).
20. Smolinske, S. C. *CRC Handbook of Food, Drug, and Cosmetic Excipients.* (CRC Press, 1992).
21. Chowhan, Z. T. Role of binders in moisture-induced hardness increase in compressed tablets and its effect on in vitro disintegration and dissolution. *J. Pharm. Sci.* **69,** 1–4 (1980).
22. Yu, L. X. & Woodcock, J. FDA pharmaceutical quality oversight. *Int. J. Pharm.* **491,** 2–7 (2015).
23. Bogdanich, W. & Hooker, J. From China to Panama, a trail of poisoned medicine. *New York Times* 2007-05-06.
24. Lukulay, P. *Pharmaceutical Manufacturing in LMICs: Where Things Go Wrong.* USP Quality Matters. 2016-05-24.
25. The Pathology of Negligence. Report of the Judicial Inquiry Tribunal to Determine the Causes of Deaths of Patients of the Punjab Institute of Cardiology, Lahore in 2011–2012. (2012).

26. Harris, G. Medicines made in India set off safety worries. *New York Times* 2014-02-14.
27. Dey, S. If I follow US standards, I will have to shut almost all drug facilities: G N Singh. *Business Standard* 2014-01-30.
28. Krishnan, V. There is no way out for Indian pharma, but to play by the world's rules. *Quartz India* 2014-06-24.
29. McNeil Jr., D. G Indian company offers to supply AIDS Drugs at low cost in Africa. *New York Times* 2001-02-07.
30. Bagri, N. T. F.D.A. Commissioner pledges more coordination with Indian regulators. *New York Times* 2014-02-18.
31. Personal communication with Kenya Pharmacy and Poisons board.
32. Government of Pakistan. *The Drugs Act, 1976*. (Pakistan Medical and Dental Council, 1976).
33. Loopholes detected in framework: Drug regulatory authority. *Dawn* 2006-10-17.
34. Junaidi, I. Appointment of Drap CEO challenged. *Dawn* 2016-10-14.
35. Lee, M. & Hirschler, B. Special Report: China's wild east drug store. *Reuters* 2012-08-28.
36. Bate, R. & Mooney, L. *Dangerous substandard medicines*. American Enterprise Institute. 2011-07-06.
37. Kingsley, P. Fake animal drugs threaten African livestock and livelihoods. *The Conversation* 2015-02-05.
38. Teko-Agbo, A., Assoumy A. & Niang Lacomev, E. M. Counterfeit drugs: Experience of West Africa. in *Regional Seminar for OIE National Focal Points for Veterinary Products—4th Cycle* (2015).

Chapter 3

1. Raustiala, K. Gin and tonic kept the British Empire healthy: The drink's quinine powder was vital for stopping the spread of malaria. *Slate* (08-2013).
2. Penn, R. The state control of medicines: The first 3000 years. *Br. J. Clin. Pharmacol.* **8,** 293–305 (1979).
3. Mann, R. D. From mithridatium to modern medicine: The management of drug safety. **81,** 725–728 (1988).
4. Sheng-Ji, P. Ethnobotanical approaches of traditional medicine studies: Some experiences from Asia. *Pharm. Biol.* **39,** 74–79 (2001).

5. World Health Organization. *National policy on traditional medicine and regulation of herbal medicines: Report of a WHO global survey.* (World Health Organization, 2005).
6. Newton, P. N., Green, M. D., Fernández, F. M., Day, N. P. & White, N. J. Counterfeit anti-infective drugs. *Lancet Infectious Diseases* **6,** 602–613 (2006).
7. Musgrave, T. & Musgrave, W. *An Empire of Plants* (Cassell and Co. UK, 2000).
8. Shah, S. *The fever: How malaria has ruled humankind for 500,000 years.* Picador, 2011.
9. Saunders, W. *Letter from Dr Edward Rigby, Norwich, 8th September 1782. Observations on the superior efficacy of the red Peruvian bark.* (1782).
10. Missouri Botanic Garden. *Proceedings of the celebration of the 300th anniversary of the first recognised use of cinchona.* 29–138 (1931).
11. Moore, E. Tests of adulterated quinine. *Lancet* 1092 (1829).
12. Croft, C. J. Adulteration of quinine. *Lancet* **31,** 292 (1838).
13. Barton, P. The great quinine fraud: Legality issues in the non-narcotic drug trade in British India. *Soc. Hist. Alcohol Drugs* **22,** (2007).
14. US FDA. *Questions and answers about FDA's enforcement action against unapproved quinine products.* (2007).
15. Fake malaria drug on the market—FDA warns. *Ghana Chronicle* 2013-12-02.
16. Journal für praktische Chemie / herausgegeben von Otto Linné Erdmann, . . . und Franz Wilhelm Schweigger-Seidel, . . . (1907).
17. Severyn, K. M. Book Review. *J. Pharm. Law* **4,** (1994).
18. Ballentine, C. *Taste of Raspberries, Taste of Death.* FDA Consumer Magazine, June, (1981).
19. Richards, I. S. & Bourgeois, M. M. In *Principles and Practice of Toxicology in Public Health* (Jones & Bartlett Learning, 2014).
20. Bate, R. *Phake: The Deadly World of Falsified and Substandard Medicines.* AEI Press, (2012).
21. FDA. *A History of the FDA and Drug Regulation in the United States.* (2006).
22. O'Brien, K. L. et al. Epidemic of pediatric deaths from acute renal failure caused by diethylene glycol poisoning. *JAMA* **279,** 1175 (1998).
23. Pandya, S. K. Letter from Bombay. An unmitigated tragedy. *BMJ* **297,** 117–119 (1988).

24. Singh, J. et al. Diethylene glycol poisoning in Gurgaon, India, 1998. *Bull. World Health Organ.* **79,** 88–95 (2001).
25. Bogdanich, W. & Hooker, J. From China to Panama, a trail of poisoned medicine. *New York Times* 2007-05-06.
26. Hanif, M. et al. Fatal renal failure caused by diethylene glycol in paracetamol elixir: The Bangladesh epidemic. *BMJ* **311,** 88–91 (1995).
27. Junod, S. W. Diethylene glycol deaths in Haiti. *Public Health Rep.* **115,** 78–86 (2000).
28. CDC. *Fatalities Associated with Ingestion of Diethylene Glycol-Contaminated Glycerin Used to Manufacture Acetaminophen Syrup—Haiti, November 1995–June 1996.* (1996).
29. Schier, J. G., Rubin, C. S., Miller, D., Barr, D. & McGeehin, M. A. Medication-associated diethylene glycol mass poisoning: A review and discussion on the origin of contamination. *J. Public Health Policy* **30,** 127–143 (2009).
30. Akuse, R. M. et al. Diagnosing renal failure due to diethylene glycol in children in a resource-constrained setting. *Pediatr. Nephrol.* **27,** 1021–1028 (2012).
31. Polgreen, L. 84 children are killed by medicine in Nigeria. *New York Times* 2009-02-06.

Chapter 4

1. Penn, R. The state control of medicines: The first 3000 years. *Br. J. Clin. Pharmacol.* **8,** 293–305 (1979).
2. Carrick, P. in 69–96 (Springer Netherlands, 1995). doi:10.1007/978-94-009-5235-5_5.
3. Butcher, S. H. (Translator) & Lang, A. (Translator). *The Odyssey of Homer: Done into English Prose.* (Macmillan, 1890).
4. De Pasquale, A. Pharmacognosy: The oldest modern science. *J. Ethnopharmacol.* **11,** 1–16 (1984).
5. Oldfather, C. H. Diodorus of Sicily, Vol 1. (1933).
6. Gwilt, J. R. Biblical ills and remedies. *J. R. Soc. Med.* **79,** 738–741 (1986).
7. Levey, M. Fourteenth-century Muslim medicine and the Hisba. *Med. Hist.* **7,** 176–182 (1963).
8. M., Ali b. in *Das Kitb "Adab ed-dunji wa"ddtn' Miwerdis* (ed. Enger, R.) (1853).

9. Al-Ghazali, *Ihya"ulum Al-din: The Revival of the Religious Sciences.* (Islamic Book Trust, 2015).
10. Sarton, G. & Siegel, F. Fifty-Seventh Critical Bibliography of the History and Philosophy of Science and of the History of Civilization (to May 1939—with special reference to cent. I to VII inclusive). (1939).
11. Levey, M. in *Texts and Documents* (1932).
12. al-ʿIbādī, Ḥunayn ibn Isḥāq Abū Zayd & Meyerhof, M. *The Book of the ten treatises on the eye ascribed to Hunain Ibn Is-Hâq (809–877 AD): The earliest existing systematic text-book of ophthalmology: the arabic text edited from the only two known manuscripts, with an English translation and glossary.* (Government Press, 1928).
13. Shanks, N. J. & Al-Kalai, D. Arabian medicine in the middle ages. *J. R. Soc. Med.* 77, 60–65 (1984).
14. Hamarneh, S. The rise of professional pharmacy in Islam. *Med. Hist.* **6,** 59–66 (1962).
15. Mann, R. D. In *International Medicines Regulations* 3–18 (Springer Netherlands, 1989). doi:10.1007/978-94-009-0857-4_1
16. Al-Baytar, I. Traité des Simples par Ibn el-Beiihar. *Trad. Lucien Leclerc, I-III. Not. Extr. des Manuscrits la Bibliothèque Natl. Paris* **23,** 25 (1881).
17. Mann, R. D. From mithridatium to modern medicine: The management of drug safety. **81,** 725–728 (1988).
18. Allbutt, R. H. S. T. C. Greek medicine in Rome. *South. Med. J.* **14,** 1015 (1921).
19. Mez-Mangold, L. History of drugs. F. Hoffman-LaRoche & Co. Ltd; 1st Am. ed. edition (1971).
20. Withington, E. T. *Medical history from the earliest times: A popular history of the healing art.* (Scientific Press, Limited, 1894).
21. O'Malley, C. D. John Evelyn and medicine. *Med. Hist.* **12,** 219–231 (1968).
22. Karaberopoulos, D., Karamanou, M. & Androutsos, G. The theriac in antiquity. *Lancet* **379,** 1942–1943 (2012).
23. Barrett, C. R. *The history of the Society of Apothecaries of London.* (E. Stock, 1905).
24. Whittet, T. D. Pepperers, spicers and grocers—forerunners of the apothecaries. *Proc. Royal Soc. Med,* **61**(8), 801–806 (1968).
25. Riley, H. T. Memorials of London life in the XIII, XIV and XV centuries. Longmans, Green and Co. (1868).

26. Trease, G. E. *Pharmacy in history.* **784,** (Baillière, Tindall and Cox, 1964).
27. Clark, G. N. & Briggs, A. *A history of the Royal College of Physicians of London.* **4,** (Oxford University Press, 2005).
28. Barrett, C. R. B. *The history of the Society of Apothecaries of London.* (E. Stock, 1905).
29. Griffin, J. P. Venetian treacle and the foundation of medicines regulation. *Br. J. Clin. Pharmacol.* 58(3), 317–325 (2004). doi: 10.1111/j.1365-2125.2004.02147.x PMCID: PMC1884566.
30. Higby, G. Apothecaries of London. *Pharmacy in History,* **24**(1), 53–54 (1982).
31. Whittet, T. D. The apothecary in provincial gilds. *Med. Hist.* **8,** 245–273 (1964).
32. Heymans, C. Pharmacology in old and modern medicine. *Annu. Rev. Pharmacol.* **7,** 1–15 (1967).
33. Urdang, G. *Pharmacopoeia londinensis of 1618.* (State Historical Society of Wisconsin, 1944).
34. Brockbank, W. Sovereign remedies: A critical depreciation of the 17th-century London Pharmacopoeia. *Med. Hist.* **8,** 1–14 (1964).
35. Dunlop, D. M. & Denston, T. C. The history and development of the "British Pharmacopoeia." *Br. Med. J.* **2,** 1250 (1958).
36. USPharmacopeia.USPmilestones—Atimeline.(2016).Availableat:http://www.usp.org/about/history-information-center/usp-milestones-timeline.
37. Janssen, W. F. Outline of the history of US drug regulation and labeling. *Food Drug Cosm. LJ* **36,** 420 (1981).
38. Boussel, P., Bonnemain, H. & Bové, F. J. *History of pharmacy and pharmaceutical industry.* **1982,** (Asklepios Press, 1983).
39. Chandler, A. D. *Shaping the industrial century: The remarkable story of the evolution of the modern chemical and pharmaceutical industries.* **46,** (Harvard University Press, 2009).
40. Drews, J. Drug discovery: A historical perspective. *Science (80-.).* **287,** 1960–1964 (2000).
41. Bell, J. & Redwood, T. Historical sketch of the progress of pharmacy. *Good, John Mason (1796). Hist. of Medicine so far as it Relates to Prof. Pharmacy. Kerrison, RM (1814). An Inq. into thePresent State Med. Prof. England. p. xii* (1843).

42. Stern, A. M. & Markel, H. The history of vaccines and immunization: Familiar patterns, new challenges. *Health Aff.* **24,** 611–621 (2005).
43. *Timeline | History of Vaccines.* (College of Physicians of Philadelphia).
44. FDA. *A History of the FDA and Drug Regulation in the United States.* (2006).
45. Wiley, H. W. *The history of a crime against the food law.* (Arno Press, 1929).
46. Ballentine, C. *Taste of raspberries, taste of death.* FDA Consumer Magazine, June, (1981).
47. Kay, G. Healthy public relations: The FDA's 1930s legislative campaign. *Bull. Hist. Med.* **75,** 446–487 (2001).
48. Juhl, R. P. Prescription to over-the-counter switch: A regulatory perspective. *Clin. Ther.* **20,** C111–C117 (1998).
49. Kim, J. H. & Scialli, A. R. Thalidomide: The tragedy of birth defects and the effective treatment of disease. *Toxicol. Sci.* **122,** 1–6 (2011).
50. Chowdhury, N., Joshi, P., Patnaik, A. & Saraswathy, B. *Administrative structure & functions of drug regulatory authorities in India.* (2016).
51. Rohit, S., Nilesh, L. B., Ravikiran, K. B., Pallavi, M. C. & Pramod, V. K. The Indian pharmaceutical industry: Evolution of regulatory system and present scenario. *Int. Res. J. Pharm.* **5,** (2012).
52. Deshpande, S. W. & Gandhi, N. Drugs and Cosmetics Act, 1940 and Rules, 1945. (2012).
53. Indian *Central Drugs Standard Control Organization.* (2016).
54. Deshpande, R. Cipla. *HBS Case Study* (2003).
55. Mueller, J. M. Taking TRIPS to India—Novartis, patent law, and access to medicines. *N. Engl. J. Med.* **356,** 541–543 (2007).
56. Lofgren, H. Novartis vs. the government of India: Patents and public health. *East Asia Forum* 1 (East Asian Bureau of Economic Research, 2013).
57. India Brand Equity Foundation. *Indian Pharma Industry: Market Size, Investment, Sector Growth Report.* (2016).
58. Thomas, K. Generic Drug Maker Pleads Guilty in Federal Case. New York Times. 2013-05-13.
59. Seiter, A. & Gyansa-Lutterodt, M. Policy note: The pharmaceutical sector in Ghana. *World Bank, Washington, DC* (2009).
60. Ghana Ministry of Trade. *Sales of Goods Act.* (1962).
61. World Intellectual Property Organization. *Ghana: Food and Drugs Act, 1992.* (2016).

62. Chinwendu, O. *The fight against fake drugs by NAFDAC in Nigeria.* 44th International Course in Health Development (ICHD) September 24, 2007 – September 12, 2008. Royal Tropical Institute [KIT], 2008.
63. Li, H., Sun, H. & Richmond, F. J. The Historical Evolution of China's Drug Regulatory System ICRS Evolution of China's Drug Regulatory System Executive Summary ICRS Evolution of China's Drug Regulatory System. (2014).
64. Ho, L. S. Market reforms and China's health care system. Soc. Sci. Med. 41, 1065–1072 (1995).
65. Bogdanich, W. & Hooker, J. From China to Panama, a Trail of Poisoned Medicine. New York Times 2007-05-06.
66. Bloomberg News. Why the FDA Is Worried About Chinese Drugs in Your Medicine Cabinet. Bloomberg News 2015-10-28.
67. Eggleston, K., Ling, L., Qingyue, M., Lindelow, M. & Wagstaff, A. Health service delivery in China: a literature review. Health Econ. 17, 149–165 (2008).
68. Lee, M. & Hirschler, B. Special Report: China's wild east drug store. Reuters 2012-08-28.
69. Yardley, J. & Barboza, D. Despite warnings, China's regulators failed to stop tainted milk. New York Times 27, 2008-09-26.

Chapter 5

1. Coulon, C. The Grand Magal in Touba: A religious festival of the mouride brotherhood of Senegal. *Afr. Aff. (Lond).* **98,** 195–210 (1999).
2. Wallis, W. The Senegal brotherhood. *Financial Times* 2007-08-29.
3. Ligne Directe Senegal. Vente illicite de médicaments à Touba—Une activité en pleine expansion—Ligne Directe, site d'informations. (2016).
4. Faucon, B., Murphy, C. & Whalen, J. Africa's malaria battle: Fake drug pipeline undercuts progress. *Wall Street Journal* 2013-05-29.
5. Ndurya, M. & Kilonzo, E. Kenya a top counterfeits market. *Daily Nation* 2015-04-07.
6. International Monetary Fund. *Kenya: Poverty Reduction Strategy Paper.* IMF Country Report No. 10/224 (2010).
7. Drugs, R. O. F., To, G., Of, S. & Board, P. *Republic of Kenya: Guidelines for Registration of Drugs.* 1–94 (2010).

8. Kangwana, B. B. et al. Malaria drug shortages in Kenya: A major failure to provide access to effective treatment. *Am. J. Trop. Med. Hyg.* **80,** 737–738 (2009).
9. Bate, R., Coticelli, P., Tren, R. & Attaran, A. Antimalarial drug quality in the most severely malarious parts of Africa—a six country study. *PLoS One* **3,** e2132 (2008).
10. Kalra, A. Sterilization deaths expose India's struggle with faulty drugs. *Reuters* 2014-11-14.
11. Kenya offers a gateway to East African boom. *Biz Community South Africa* 2013-11-12.
12. Uganda Radio Network. Stolen govt drugs sold in Kenya. *Uganda Radio Network* 2012-10-16.
13. OECD. *Global Forum on Competition: Competition Issues in the Distribution of Pharmaceuticals; Contribution from Indonesia.* (2014).
14. Zulkifli, N. W., Aziz, N. A., Hassan, Y., Hassali, M. A. & Bahrin, N. L. Z. The development of the problem solving framework in managing unregistered drugs: Pharmacists perspectives. *J. Pharm. Pract. Community Med.* **2,** (2016).
15. Holloway, K. A. Pharmaceuticals in health care delivery. *Mission Rep.* **30,** (2011).
16. Testimony—Securing the pharmaceutical supply chain. Hearing of the committee on health, education, labor and pensions. US Senate. One hundred Twelfth congress, First session. 2011-09-14. (2011).
17. Bogdanich, W. & Hooker, J. From China to Panama, a trail of poisoned medicine. *New York Times* 2007-05-06.
18. Government of Pakistan. *The Drugs Act, 1976.* (Pakistan Medical and Dental Council, 1976).
19. Nishtar, S. Choked pipes. Oxford University Press, (2010).
20. Nishtar, S. et al. Health reform in Pakistan: A call to action. *Lancet* **381,** 2291–2297 (2013).
21. Nishtar, S. et al. Pakistan's health system: Performance and prospects after the 18th Constitutional Amendment. *Lancet* **381,** 2193–2206 (2013).
22. *DRAP Act, Government of Pakistan.* (2012).
23. Rashid, H. Impact of the Drug Regulatory Authority, Pakistan: An Evaluation. *New Visions Public Aff.* **7,** (2015).

24. Zaidi, S., Bigdeli, M., Aleem, N. & Rashidian, A. Access to essential medicines in Pakistan: Policy and health systems research concerns. *PLoS One* **8,** e63515 (2013).
25. Somra, G. The backstreet labs feeding Pakistan's fake drug trade. *CNN* 2015-08-30.
26. Pant M., Pande D. (2017) India–Pakistan Trade: An Analysis of the Pharmaceutical Sector. In Taneja N., Dayal I. (eds) *India-Pakistan Trade Normalisation.* Springer, Singapore.
27. Aftab, M. New trade policy to help boost Pakistan exports. *Khaleej Times* 2016-04-04.
28. Junaidi, I. No law to stop pharma companies from importing raw materials. *Dawn* 2015-07-15.
29. PPMA calls for deregulation of DRAP. *Daily Times* 2016-11-15.
30. Adnan, I. Following safety protocols: Seeking standards, Punjab to set up modern testing lab. *Express Tribune* 2017-01-16.
31. Sahoo, N., Manchikanti, P. & Dey, S. Herbal drugs: Standards and regulation. *Fitoterapia* **81,** 462–471 (2010).
32. Harris, G. U.S. identifies tainted Heparin in 11 countries. *New York Times* 2008-04-22.
33. Eichenwald, K. Killer pharmacy: Inside a medical mass murder case. *Newsweek* 2015-04-16.
34. Bate, R. *Phake: The Deadly World of Falsified and Substandard Medicines.* AEI Press, (2012).
35. Newton, P. N. et al. The primacy of public health considerations in defining poor quality medicines. *PLoS Med.* **8,** e1001139 (2011).
36. Mueller, J. M. The tiger awakens: The tumultuous transformation of India's patent system and the rise of Indian pharmaceutical innovation. *U. Pitt. L. Rev.* **68,** 491 (2006).
37. *IMPACT: International Medical Products Anti-Counterfeiting Taskforce.* (2006).
38. Clift, C. *Combating Counterfeit, Falsified and Substandard Medicines: Defining the Way Forward?* Chatham House Briefing Paper, November, (2010).
39. Baker, B. K. Settlement of India/EU WTO dispute reseizures of in-transit medicines: Why the proposed EU border regulation isn't good enough. PIJIP Research Paper no. 2012-02 American University Washington College of Law, Washington, D.C. (2012).

40. Shukla, N. & Sangal, T. Generic drug industry in India: The counterfeit spin. *Journal of Intellectual Property Rights*, **14**, 236–240 (2009).
41. Burci, G. L. Public health and "counterfeit" medicines: The role of the World Health Organization. American Society of International Law, ASIL Insights, 2013-01-11.
42. Counterfeit drugs on rise, pose global threat—WHO. *Reuters* 2010-05-19.
43. Ossola, A. The fake drug industry is exploding, and we can't do anything about it. *Newsweek* 2015-09-17.
44. Deshpande, R. Cipla. *HBS Case Study* (2003).

Chapter 6

1. Mitchell, K. Pakistani man sentenced for smuggling counterfeit drugs into the U.S. *Denver Post* 2016-10-18.
2. *US FDA: "Counterfeit Drugs: Fighting Illegal Supply Chains"* (Office of the Commissioner, 2014).
3. Lewis, K. The fake and the fatal: The consequences of counterfeits. *Park Place Econ.* **17,** 14 (2009).
4. Blackstone, E. A., Fuhr, J. P. & Pociask, S. The health and economic effects of counterfeit drugs. *Am. Heal. Drug Benefits* **7,** 216–224 (2014).
5. Moken, M. C. Fake pharmaceuticals: How they and relevant legislation or lack thereof contribute to consistently high and increasing drug prices. *Am. JL Med.* **29,** 525 (2003).
6. Stearn, D. W. Deterring the importation of counterfeit pharmaceutical products. *Food Drug LJ* **59,** 537 (2004).
7. The Pathology of Negligence. Report of the Judicial Inquiry Tribunal to Determine the Causes of Deaths of Patients of the Punjab Institute of Cardiology, Lahore in 2011–2012. (2012).
8. Supreme Court disposes of case against Efroze Chemical. *Pakistan Business Recorder* 2014-03-20.
9. SC asks pharma firm to pay compensation in six months. *Dawn* 2013-12-31.
10. Ames, J. & Souza, D. Z. Falsificação de medicamentos no Brasil. *Rev. Saude Publica* **46,** 154–159 (2012).
11. Gautam, C. S., Utreja, A. & Singal, G. L. Spurious and counterfeit drugs: A growing industry in the developing world. *Postgrad. Med. J.* **85,** 251–256 (2009).

12. Erhun, W. O., Babalola, O. O. & Erhun, M. O. Drug regulation and control in Nigeria: The challenge of counterfeit drugs. *J. Health Popul. Dev. Ctries.* **4,** 23–34 (2001).
13. Bate, R. & Nugent, R. The deadly world of fake drugs. *Foreign Policy* 57 (2008).
14. Ossola, A. The fake drug industry is exploding, and we can't do anything about it. *Newsweek* 2015-09-17.
15. Marete, G. Terrorists cash in on counterfeit medicine. *Daily Nation* 2016-09-01.
16. Oketch, A. Why devolved health care is a bitter pill to swallow for some counties. *Daily Nation* 2014-04-10.
17. Council of Europe. *The Medicrime Convention.* (2011).
18. Keitel, S. The Medicrime convention: Criminalising the falsification of medicines and similar crimes. *GBI J* **1,** 138–141 (2012).
19. Council of Europe. Medicrime Convention Treaty 211. (2011).
20. Attaran, A. & Bate, R. A Counterfeit drug treaty: Great idea, wrong implementation. *Lancet* **376,** (2010).
21. Mackey, T. K. & Liang, B. A. The global counterfeit drug trade: Patient safety and public health risks. *J. Pharm. Sci.* **100,** 4571–4579 (2011).
22. Interpol. *Pharmaceutical crime.* (2014).
23. Interpol. Operation Pangea. *Interpol Reports* (2015). Available at: https://www.interpol.int/Crime-areas/Pharmaceutical-crime/Operations/Operation-Pangea. (Accessed: 11th February 2017)
24. Interpol. *Illegal online medicine suppliers targeted in first international Internet day of action.* 2008-11-13.
25. Interpol. *International operation combats online supply of counterfeit and illegal medicines.* 2009-11-20.
26. Interpol. *Online sale of fake medicines and products targeted in INTERPOL operation.* 2016-07-10.
27. Interpol. *Operation Mamba.* (2010).
28. Interpol. *Operation Cobra.* (2011).
29. Interpol. *Operation Porcupine.* (2014).
30. Interpol. Operation Heera. 2017.
31. *Operation Giboia. Interpol Reports* (2015).
32. Interpol. Operation Storm. *INTERPOL Reports* (2015). Available at: https://www.interpol.int/Crime-areas/Pharmaceutical-crime/Operations/Operation-Storm. (Accessed: 11th February 2017)

33. Mudur, G. India to introduce death penalty for peddling fake drugs. *BMJ Br. Med. J.* **327,** 414 (2003).
34. Lewis, K. China's counterfeit medicine trade booming. *CMAJ* **181,** E237–E238 (2009).
35. Wertheimer, A. I. & Wang, P. G. *Counterfeit medicines; v.1: Policy, economics and countermeasures.* ILM Publications (2012), p. 57.
36. Dasgupta, S. Death for 6 Chinese traders faking "Made in India" tags. *Times of India* 2009-12-10.
37. Kahn, J. China quick to execute drug official. *New York Times* 2007-07-11.

Chapter 7

1. Luo, W. & Najdawi, M. Trust-building measures: A review of consumer health portals. *Commun. ACM* **47,** 108–113 (2004).
2. Hopkins, D. M., Kontnik, L. T. & Turnage, M. T. *Counterfeiting exposed: Protecting your brand and customers.* (J. Wiley & Sons, 2003).
3. Kessel, M. Restoring the pharmaceutical industry's reputation. *Nat Biotech* **32,** 983–990 (2014).
4. Bate, R. *Phake: The Deadly World of Falsified and Substandard Medicines.* AEI Press, (2012).
5. Smedley, T. Big pharma attempts to cast off bad reputation by targeting the poor. *Guardian* 2015-06-25.
6. Lofgren, H. Novartis vs. the government of India: Patents and public health. in *East Asia forum* 1 (East Asian Bureau of Economic Research, 2013).
7. Mueller, J. M. Taking TRIPS to India—Novartis, patent law, and access to medicines. *N. Engl. J. Med.* **356,** 541–543 (2007).
8. Sampat, B. N., Shadlen, K. C. & Amin, T. M. Challenges to India's pharmaceutical patent laws. *Science (80-.).* **337,** 414–415 (2012).
9. 't Hoen, E. A victory for global public health in the Indian Supreme Court. *J. Public Health Policy* **34,** 370–374 (2013).
10. Baker, J. R. EpiPen pricing controversy reflects larger issues in pharma industry. *Stat News* 2016-09-28.
11. McCarthy, M. Drug's 5000% price rise puts spotlight on soaring US drug costs. *BMJ Br. Med. J.* **351,** (2015).
12. Schoonveld, M. E. *The price of global health: Drug pricing strategies to balance patient access and the funding of innovation.* (Gower Publishing, Ltd., 2015).

13. Fahsi, M. Medical neocolonialism: Big pharma outsources unethical clinical trials to South Africa. *MintPress News* 2013-07-22.
14. Nundy, S. & Gulhati, C. M. A new colonialism?—Conducting clinical trials in India. *N. Engl. J. Med.* **352,** 1633–1636 (2005).
15. Snell, B. Inappropriate drug donations: The need for reforms. *Lancet* **358,** 578–580 (2001).
16. Pinheiro, C. P. Drug donations: What lies beneath. *Bull. World Health Organ.* **86,** 580A–580A (2008).
17. Shah, A. Pharmaceutical corporations and medical research. *Global Issues.* 2010-10-02.
18. Lyons, T. Globalisation, failed states and pharmaceutical colonialism in Africa. *Australas. Rev. African Stud.* **30,** 68 (2009).
19. Lyons, T. J. Pharmaceutical colonialism—Ethical issues for research in Africa. International Conference Proceedings of the Annual AFSAAP Conference (2009).
20. Gilson, L. Trust and the development of health care as a social institution. *Soc. Sci. Med.* **56,** 1453–1468 (2003).
21. Rothstein, B. & Uslaner, E. M. All for all: Equality, corruption, and social trust. *World Polit.* **58,** 41–72 (2005).
22. Waning, B., Diedrichsen, E. & Moon, S. A lifeline to treatment: The role of Indian generic manufacturers in supplying antiretroviral medicines to developing countries. *J. Int. AIDS Soc.* **13,** 35 (2010).
23. Jack, A. The man who battled big pharma. *Financial Times* 2008-05-29.
24. Deshpande, R. Cipla. *HBS Case Study* (2003).
25. IFPMA. *IFPMA in Brief.* (2016).
26. IFPMA. *Improving global health through collaborations and dialogue representing the research-based pharmaceutical industry.* (2016).
27. IFPMA. *Code of practice.* (2012).
28. Personal communication with Pfizer, 2013.
29. Grabowski, H. Patents, innovation and access to new pharmaceuticals. *J. Int. Econ. Law* **5,** 849–860 (2002).
30. Gervais, D. J. *The TRIPS agreement: Drafting history and analysis.* (Sweet & Maxwell, 2003).
31. World Health Organization. Globalization, TRIPS and access to pharmaceuticals. (2001).
32. Li, L. Technology designed to combat fakes in the global supply chain. *Bus. Horiz.* **56,** 167–177 (2013).

33. Paul, S. M. et al. How to improve R&D productivity: The pharmaceutical industry's grand challenge. *Nat. Rev. Drug Discov.* **9,** 203–214 (2010).
34. Hall, A. & Antonopoulos, G. A. in *Fake Meds Online* 1–17 (Springer, 2016).
35. Harris, G. Medicines made in India set off safety worries. *NY Times* 2014-02-14.
36. Somra, G. Online pharmacies suspected of counterfeit drug sales. *CNN* 2015-08-31.
37. Blackstone, E. A., Fuhr, J. P. & Pociask, S. The health and economic effects of counterfeit drugs. *Am. Heal. Drug Benefits* **7,** 216–224 (2014).
38. Merck Inc. *Merck for mothers Program.* http://merckformothers.com/.
39. GSK health for all Initiative. https://www.gsk.com/en-gb/responsibility/health-for-all/
40. Novartis Access Program. https://www.novartis.com/about-us/corporate-responsibility/expanding-access-healthcare/novartis-social-business/novartis-access
41. IFPMA. *Regulatory conferences of IFPMA: Strengthening regulatory systems.* (2016).
42. Shukla, N. & Sangal, T. Generic drug industry in India: The counterfeit spin. *Journal of Intellectual Property Rights,* 14, 236–240, 2009.
43. Pearl, D. & Freedman, A. Altruism, politics and bottom line intersect at Indian generics firm. *Wall Street Journal* 2001-03-12.
44. Haley, G. T. & Haley, U. C. V. The effects of patent-law changes on innovation: The case of India's pharmaceutical industry. *Technol. Forecast. Soc. Change* **79,** (2012).
45. Cipla aims for export growth. *Manufacturing Chemist Pharma* 2014-04-04.
46. Cipla News. *Cipla to launch South Africa's first biotech manufacturing facility.* 2016-07-08.
47. Dey, S. If I follow US standards, I will have to shut almost all drug facilities: G N Singh. *Business Standard* 2014-01-30.
48. Zaman, M. H. Private communication with CEO, Ferozesons Pharmaceuticals, Lahore. (2016).
49. Hussain, D. Tangled case of drug pricing. *Dawn* (2014).
50. Rasmussen, S. E. Killing, not curing: Deadly boom in counterfeit medicine in Afghanistan. *Guardian* 2015-01-07.
51. Brhlikova, P. et al. Trust and the regulation of pharmaceuticals: South Asia in a globalised world. *Global. Health* **7,** 10 (2011).

Chapter 8

1. Wellcome Trust Press Release. *Counterfeit and substandard antimalaria drugs threaten crisis in Africa, experts warn*. 2012-01-16.
2. Nayyar, G. M. L., Breman, J. G., Newton, P. N. & Herrington, J. Poor-quality antimalarial drugs in southeast Asia and sub-Saharan Africa. *Lancet. Infect. Dis.* **12,** 488–96 (2012).
3. Tun, K. M. *et al.* Spread of artemisinin-resistant Plasmodium falciparum in Myanmar: A cross-sectional survey of the K13 molecular marker. *Lancet Infect. Dis.* **15,** 415–421 (2015).
4. Newton, P. Investment is key to tackling the ongoing threat of fake medicines. *Oxford Science Blog* 2016-12-13.
5. Williams, L. & McKnight, E. The real impact of counterfeit medications. *US Pharm.* **39,** 44–46 (2014).
6. Jha, P. Reliable direct measurement of causes of death in low- and middle-income countries. *BMC Med.* **12,** 19 (2014).
7. Measuring mortality in developing countries. *PLoS Med.* **3,** e56 (2005).
8. Ahmed, Z. Why the census threatens the political status quo in Pakistan. *The Nation* 2016-04-02.
9. Burton, E. C. & Collins, K. A. Religions and the autopsy. *Medscape News Perspect.* 2012-03-20.
10. Bassat, Q. *et al.* Development of a post-mortem procedure to reduce the uncertainty regarding causes of death in developing countries. *Lancet Glob. Heal.* **1,** e125–e126 (2013).
11. Cohen, J. M. *et al.* Optimizing investments in malaria treatment and diagnosis. *Science (80-.).* **338,** 612–614 (2012).
12. Kunin, C. M., Johansen, K. S., Worning, A. M. & Daschner, F. D. Report of a symposium on use and abuse of antibiotics worldwide. *Rev. Infect. Dis.* **12,** 12–19 (1990).
13. Kelesidis, T. & Falagas, M. E. Substandard/counterfeit antimicrobial drugs. *Clin. Microbiol. Rev.* **28,** 443–464 (2015).
14. Buckley, G. J. & Gostin, L. O. The effects of falsified and substandard drugs. (2013).
15. Renschler, J. P., Walters, K., Newton, P. N. & Laxminarayan, R. Estimated under-five deaths associated with poor-quality antimalarials in sub-Saharan Africa. *Am. J. Trop. Med. Hyg.* 14–725 (2015).

16. Sibley, C. H., Barnes, K. I., Watkins, W. M. & Plowe, C. V. A network to monitor antimalarial drug resistance: A plan for moving forward. *Trends Parasitol.* **24,** 43–48 (2008).
17. Blower, S. M. & Dowlatabadi, H. Sensitivity and uncertainty analysis of complex models of disease transmission: An HIV model, as an example. *Int. Stat. Rev. Int. Stat.* 229–243 (1994).
18. Bird, R. C. Counterfeit drugs: A global consumer perspective. *Wake For. Intell. Prop. LJ* **8,** 387 (2007).
19. Ashley, E. A. *et al.* Spread of artemisinin resistance in Plasmodium falciparum malaria. *N. Engl. J. Med.* **371,** 411–423 (2014).
20. Azhar, S. *et al.* The role of pharmacists in developing countries: The current scenario in Pakistan. *Hum. Resour. Health* **7,** 54 (2009).
21. Laing, R. O., Hogerzeil, H. V & Ross-Degnan, D. Ten recommendations to improve use of medicines in developing countries. *Health Policy Plan.* **16,** 13–20 (2001).
22. Erhun, W. O., Babalola, O. O. & Erhun, M. O. Drug regulation and control in Nigeria: The challenge of counterfeit drugs. *J. Health Popul. Dev. Ctries.* **4,** 23–34 (2001).
23. Ghayur, M. N. Pharmacy education in developing countries: Need for a change. *Am. J. Pharm. Educ.* **72,** (2008).
24. Pharmacy Council Pakistan. *Pharm D. Curriculum in Pakistan.* Revised 2013.
25. Kwame Nkrumah University of Science and Technology. Faculty of Pharmacy and Pharmaceutical Sciences, KNUST—Pharm D Course Outline. Revised 2016.
26. Zofou, D. *et al.* The needs of biomedical science training in Africa: Perspectives from the experience of young scientists. *African J. Heal. Prof. Educ.* **3,** 9–12 (2011).
27. Perry, L. & Malkin, R. Effectiveness of medical equipment donations to improve health systems: How much medical equipment is broken in the developing world? *Med. Biol. Eng. Comput.* **49,** (2011).
28. Harris, E. Building scientific capacity in developing countries. *EMBO Rep.* **5,** 7–11 (2004).
29. Toklu, H. Z. & Hussain, A. The changing face of pharmacy practice and the need for a new model of pharmacy education. *J. Young Pharm.* **5,** 38–40 (2013).
30. Bunoti, S. The quality of higher education in developing countries needs professional support. In *22nd International Conference on Higher*

Education. Retrieved from http://www.intconfhighered.org/FINAL%20 Sarah%20Bunoti.pdf (2011).
31. Woolman, D. C. Educational reconstruction and post-colonial curriculum development: A comparative study of four African countries. *Int. Educ. J.* **2,** 27–46 (2001).
32. Zaman, M. H. Ethiopian longhorns and our curriculum. *Express Tribune* 2013-11-10.
33. Dambudzo, I. I. Curriculum Issues: Teaching and learning for sustainable development in developing countries: Zimbabwe case study. *J. Educ. Learn.* **4,** 11 (2015).
34. Kassi, M. Pharmacy in Africa. Ohio State University, College of Pharmacy, 2016.
35. Anderson, C. *et al.* Global perspectives of pharmacy education and practice. *World Med. Heal. Policy* **2,** 5–18 (2010).
36. Kheir, N. *et al.* Pharmacy education and practice in 13 Middle Eastern countries. *Am. J. Pharm. Educ.* **72,** 133 (2008).
37. Smesny, A. L. *et al.* Barriers to scholarship in dentistry, medicine, nursing, and pharmacy practice faculty. *Am. J. Pharm. Educ.* **71,** 91 (2007).
38. Bililign, S. The need for interdisciplinary research and education for sustainable human development to deal with global challenges. *Int. J. African Dev.* **1,** 8 (2013).
39. Wall, K. *Engineering: Issues, challenges and opportunities for development.* (UNESCO, 2010).
40. Lustick, D. R. & Zaman, M. H. Biomedical engineering education and practice challenges and opportunities in improving health in developing countries. In *2011 Atlanta Conference on Science and Innovation Policy* 1–5 (2011). doi:10.1109/ACSIP.2011.6064477
41. Khalid, S. M. & Khan, M. F. Pakistan: The state of education. *Muslim World* **96,** 305–322 (2006).
42. Zaman, M. H. Why choose between bio and comp sci? *Express Tribune* 2012-01-17.
43. Pakistan Medical and Dental Council. *MBBS Curriculum Pakistan.*
44. Lam, T. P. & Lam, Y. Y. B. Medical education reform: The Asian experience. *Acad. Med.* **84,** 1313–1317 (2009).
45. Nagaraj, A. *et al.* Counterfeit medication: Perception of doctors and medical wholesale distributors in western India. *J. Int. Soc. Prev. Community Dent.* **5,** S7 (2015).

46. Okeke, I. N., Lamikanra, A. & Edelman, R. Socioeconomic and behavioral factors leading to acquired bacterial resistance to antibiotics in developing countries. *Emerg. Infect. Dis.* **5,** 18–27 (1999).
47. Torloni, M. R., Gomes Freitas, C., Kartoglu, U. H., Metin Gülmezoglu, A. & Widmer, M. Quality of oxytocin available in low and middle income countries: A systematic review of the literature. *BJOG An Int. J. Obstet. Gynaecol.* **123,** 2076–2086 (2016).
48. Personal communication with Jhpiego field offices in India, and with ob/gyn physicians in Nepal. (2016).
49. Alfa-Wali, M., Mohammed, I. & Yusuph, H. Under the counter, underground or unethical medicines? An African perspective. *Trop. Doct.* 49475513512639 (2013).
50. Bansal, R. K. & Das, S. Unethical relationship between doctors and drugs companies. *JIAFM,* **27**(1), 40–42 (2005).
51. IFPMA. IFPMA Workshop on Counterfeit Medicines, Senegal, April 2015. (2015).
52. IFPMA. 27th IFPMA Assembly 2014. (2014).
53. Mhando, L., Jande, M. B., Liwa, A., Mwita, S. & Marwa, K. J. Public awareness and identification of counterfeit drugs in Tanzania: A view on antimalarial drugs. *Adv. Public Heal.* **2016,** (2016).
54. Abdoulaye, I., Chastanier, H., Azondekon, A., Dansou, A. & Bruneton, C. Evaluation of public awareness campaigns on counterfeit medicines in Cotonou, Benin. *Med. Trop. Rev. du Corps sante Colon.* **66,** 615–618 (2006).
55. Zaman, M. H. Interviews with Samaa TV Pakistan producers and Express News, Pakistan producers. (2014).
56. Faucon, B., Murphy, C. & Whalen, J. Africa's malaria battle: Fake drug pipeline undercuts progress. *Wall Street Journal.*
57. Bogdanich, W. & Hooker, J. From China to Panama, a trail of poisoned medicine. *New York Times* 2007-05-06.
58. Punjab Information Technology Board. *Drug Inspection and Monitoring Evaluation System.* Accessed at: https://www.pitb.gov.pk/dime
59. Shearlaw, M. A cellphone is no substitute for a midwife, African tech prodigy warns. *Guardian* 2016-12-06.
60. Adepoju, P. Nigerian government announces hotlines for fake malaria drugs. *HealthNews NG* 2016-08-22.
61. Jamah, A. Kenya: Boost to war on counterfeit drugs in Kenya as agency acquires instant testing equipment. *Daily Standard* 2015-10-28.

62. Jakarta Post. BPOM to launch hotline against illegal drugs, cosmetics. *Jakarta Post* 2015-06-02.
63. Garuba, H. A., Kohler, J. C. & Huisman, A. M. Transparency in Nigeria's public pharmaceutical sector: Perceptions from policy makers. *Global. Health* **5,** 14 (2009).
64. Lancet, T. Counterfeit drugs: A growing global threat. *Lancet* **379,** 685 (2012).

Chapter 9

1. Bhadelia, N. Most medical equipment in low-income countries is donated. And most of it doesn't work. *NPR* 2016-09-08.
2. Jones, A. Medical equipment donated to developing nations usually ends up on the junk heap. *Scientific American* 2013-05-06.
3. Perry, L. & Malkin, R. Effectiveness of medical equipment donations to improve health systems: How much medical equipment is broken in the developing world? *Med. Biol. Eng. Comput.* **49,** (2011).
4. Mackey, T. K. & Liang, B. A. Improving global health governance to combat counterfeit medicines: A proposal for a UNODC-WHO-Interpol trilateral mechanism. *BMC Med.* **11,** 233 (2013).
5. Cohen, J. C., Mrazek, M. F. & Hawkins, L. Corruption and pharmaceuticals: Strengthening good governance to improve access. *Many Faces Corrupt. Track. Vulnerabilities Sect. Level. Washingt. DC World Bank* (2007).
6. Kovacs, S. et al. Technologies for detecting falsified and substandard drugs in low and middle-income countries. *PLoS One* **9,** e90601 (2014).
7. Hall, C. Technology for combating counterfeit medicine. *Pathog. Glob. Health* **106,** 73–76 (2012).
8. Karger, B. L. HPLC: Early and recent perspectives. *J. Chem. Educ* **74,** 45 (1997).
9. Henry, R. A. The early days of HPLC at DuPont. *LCGC North America,* **27**(2), 146–153 (2009).
10. PR Newswire. *Global HPLC Systems and Accessories Industry.* 2014-05-21.
11. Ahuja, S. & Dong, M. *Handbook of pharmaceutical analysis by HPLC.* **6,** (Elsevier, 2005).
12. Personal communication with FDA Ghana.
13. Personal communication with DRAP Pakistan.

14. US Pharmacopeia. *Establishing a Network of Drug Quality Control Laboratories in USAID-supported African Countries.* USAID Report. 2009-09-18.
15. Zaman, M. H. Personal communication with Kenya National Quality Control Lab. (2014).
16. Fonjungo, P. N. et al. Laboratory equipment maintenance: A critical bottleneck for strengthening health systems in sub-Saharan Africa? *J. Public Health Policy* **33,** 34–45 (2012).
17. Malkin, R. A. Design of health care technologies for the developing world. *Annu. Rev. Biomed. Eng.* **9,** 567–587 (2007).
18. Jähnke, R. W. O., Küsters, G. & Fleischer, K. Low-cost quality assurance of medicines using the GPHF-Minilab®. *Drug Inf. J.* **35,** 941–945 (2001).
19. Fernandez, F. M. et al. Poor quality drugs: Grand challenges in high throughput detection, countrywide sampling, and forensics in developing countries. *Analyst* **136,** 3073–3082 (2011).
20. Höllein, L., Kaale, E., Mwalwisi, Y. H., Schulze, M. H. & Holzgrabe, U. Routine quality control of medicines in developing countries: Analytical challenges, regulatory infrastructures and the prevalence of counterfeit medicines in Tanzania. *TrAC Trends Anal. Chem.* **76,** 60–70 (2016).
21. Hajjou, M., Qin, Y., Bradby, S., Bempong, D. & Lukulay, P. Assessment of the performance of a handheld Raman device for potential use as a screening tool in evaluating medicines quality. *J. Pharm. Biomed. Anal.* **74,** 47–55 (2013).
22. Sproxil. *Sproxil* (2008). Available at: Sproxil.com.
23. Isah, H. Information and communication technology in combating counterfeit drugs. *arXiv Prepr. arXiv1211.1242* (2012).
24. Judicial Inquiry Tribunal. The Pathology of Negligence. Report of the Judicial Inquiry Tribunal to Determine the Causes of Deaths of Patients of the Punjab Institute of Cardiology, Lahore in 2011–2012. (2012).
25. Weaver, A. A. et al. Paper analytical devices for fast field screening of beta lactam antibiotics and antituberculosis pharmaceuticals. *Anal. Chem.* **85,** 6453–6460 (2013).
26. Weaver, A. A. & Lieberman, M. Paper test cards for presumptive testing of very low quality antimalarial medications. *Am. J. Trop. Med. Hyg.* **92,** 17–23 (2015).
27. Green, M. D. Antimalarial drug resistance and the importance of drug quality monitoring. *J. Postgrad. Med.* **52,** 288 (2006).
28. Johnston, A. & Holt, D. W. Substandard drugs: A potential crisis for public health. *Br. J. Clin. Pharmacol.* **78,** 218–243 (2014).

29. Bansal, D., Malla, S., Gudala, K. & Tiwari, P. Anti-counterfeit technologies: A pharmaceutical industry perspective. *Sci. Pharm.* **81,** 1–14 (2012).
30. Cadwalladr, C. New Africa: Ghanaian tech innovator who led counterfeit drugs crackdown. *Guardian* 2012-08-25.
31. Warsi, S. A Pakistani startup is trying to banish fake prescription drugs. *Vice* 2016-04-04.
32. Ashok, I. Scientists develop $1 paper card that can detect counterfeit drugs. *International Business Times* 2016-08-23.
33. Desai, D. Pharmachk: Robust device for counterfeit and substandard medicines screening on developing regions. PhD Thesis in the Department of Biomedical Engineering, Boston University. (2014).
34. Whitesides, G. M. The origins and the future of microfluidics. *Nature* **442,** 368–373 (2006).
35. Batson, J. S. et al. Assessment of the effectiveness of the CD3+ tool to detect counterfeit and substandard anti-malarials. *Malar. J.* **15,** 119 (2016).
36. Eisberg, N. Mass spec for the masses. *Chemistry & Industry* (2012). doi:10.1002/cind.7605_17
37. Caudron, J. et al. Substandard medicines in resource-poor settings: A problem that can no longer be ignored. *Trop. Med. Int. Heal.* **13,** 1062–1072 (2008).
38. Yager, P., Domingo, G. J. & Gerdes, J. Point-of-care diagnostics for global health. *Annu. Rev. Biomed. Eng.* **10,** 107–144 (2008).
39. Idriss, M. The African startup using phones to spot counterfeit drugs. Bloomberg, 2015-07-31.
40. Clifford, K. L. & Zaman, M. H. Engineering, global health, and inclusive innovation: Focus on partnership, system strengthening, and local impact for SDGs. *Glob. Health Action* **9,** (2016).
41. Yamey, G. & Morel, C. Investing in health innovation: A cornerstone to achieving global health convergence. *PLoS Biol* **14,** e1002389 (2016).
42. Mangham, L. J. & Hanson, K. Scaling up in international health: What are the key issues? *Health Policy Plan.* **25,** 85–96 (2010).
43. McNerney, R. Diagnostics for developing countries. *Diagnostics* **5,** 200–209 (2015).
44. Lamph, S. Regulation of medical devices outside the European Union. *J. R. Soc. Med.* **105,** S12–S21 (2012).
45. Chao, T. E. & Mody, G. N. The impact of intellectual property regulation on global medical technology innovation. *BMJ Innov.* **1,** 49–50 (2015).

46. Chesbrough, H. & Rosenbloom, R. S. The role of the business model in capturing value from innovation: Evidence from Xerox Corporation's technology spin-off companies. *Ind. Corp. Chang.* **11,** 529–555 (2002).
47. Saving Lives at Birth. Saving lives at birth: A grand challenge for development. *Saving Lives at Birth.* Available at: https://savinglivesatbirth.net/problem.
48. USP. USP Fellowships | U.S. Pharmacopeial Convention.
49. Buckley, G. J. & Gostin, L. O. Causes of falsified and substandard drugs. Gillian Buckley & Lawrence O. Gostin eds., The National Academies Press 2013.

Chapter 10

1. Agbogbloshie: The world's largest e-waste dump. *Guardian* 2014-02-27.
2. Piervencenzi, R. Making progress in fighting counterfeit drugs in Africa. *Sci Dev Net* 2016-02-22.
3. Johnston, A. & Holt, D. W. Substandard drugs: A potential crisis for public health. *Br. J. Clin. Pharmacol.* **78,** 218–243 (2014).
4. Buckley, G. J. & Gostin, L. O. Causes of falsified and substandard drugs. Gillian Buckley & Lawrence O. Gostin eds., The National Academies Press 2013.
5. Cooperative Agreement USAID-USP on Promoting Quality of Medicines. PROMOTING QUALITY OF MEDICINES (PQM). (2009).
6. Heyman, M. L. & Williams, R. L. Ensuring global access to quality medicines: Role of the US pharmacopeia. *J. Pharm. Sci.* **100,** 1280–1287 (2011).
7. Management Sciences for Health. *Improving Drug Management for Public Health: Lessons from the Rational Pharmaceutical Management Project.* (2001).
8. Carpenter, Joyce. A Review of Drug Quality in 11 Asian Countries with Focus on Anti-infectives. USP Technical Report. 2004.
9. USAID Report. *Amazon Malaria Initiative: Goals and Accomplishments.* (2010).
10. Phanouvong, S. & Smine, A. *U. S. Pharmacopeia Drug Quality and Information Review and Assessment of Drug Quality Assurance and Control in Madagascar USP DQI Review and Assessment Madagascar Drug Quality Assurance and Control.* (2003).
11. Ratanawijitrasin, S. & Phanouvong, S. *The state of medicine quality in the Mekong subregion.* PQM-USAID Occasional Paper. (2014).
12. Interpol. *Pharmaceutical crime.* Available at: https://www.interpol.int/Crime-areas/Pharmaceutical-crime/Pharmaceutical-crime.

13. USP. *Promoting Quality of Medicines*. Available at: http://www.usp.org/global-health/promoting-quality-medicines.
14. Krech, L. A. *et al.* The medicines quality database: A free public resource. *Bull. World Health Organ.* **92,** 2–2A (2014).
15. Hajjou, M. *et al.* Monitoring the quality of medicines: Results from Africa, Asia, and South America. *Am. J. Trop. Med. Hyg.* **92,** 68–74 (2015).
16. USP Announces PharmaCheck Drug Quality Screening Technology Recognized Among "World Changing Ideas Of 2013." *Drug Discovery Online* 2013-12-03.
17. US Pharmacopeia. Glossary of PQM Activities. Available at: http://www.usp.org/global-health/promoting-quality-medicines.
18. Fight the Fakes. *About Fight the Fakes Campaign*. Available at: http://fightthefakes.org/.
19. Fight the Fakes. *Terms of Reference: Fight the Fakes Campaign*. (2013).
20. Fight the Fakes. *Fight the Fakes: What are we looking to achieve ? First 2 years at a glance*. Fight the Fakes Report. (2015).
21. Fight the Fakes News. *Fight the Fakes secretariat transitions to IFPW*. 2017-02-09.
22. University College London News. *UCL School of Pharmacy joins the Fight the Fakes campaign*. 2016-02-10.
23. Johnson, K. Counterfeit Drugs Around the World | The difference between fake pharmaceuticals and real ones. Available at: https://counterfeitdrugs.wordpress.com/.
24. American Society of Tropical Medicine and Hygiene News. *ASTMH Continues to Push Against Falsified Medicines*. 2015-12-09.
25. Ossola, A. The fake drug industry is exploding, and we can't do anything about it. *Newsweek* 2015-09-17.
26. Bate, R. *Phake: The Deadly World of Falsified and Substandard Medicines*. AEI Press. (2012).
27. Yanagizawa-Drott, D. *Harnessing Market Forces to Fight Fake Drugs*. Innovations for Poverty Action. 2012-10-05.
28. BBC. Tanzania ivory: China officials "went on buying spree." *BBC* 2014-11-06.
29. Vidal, J. Elephants could vanish from one of Africa's key reserves within six years. *Guardian* 2016-06-01.
30. Timmons, H. Chinese officials allegedly smuggled ivory out of Tanzania on president Xi Jinping's plane. *Quartz* 2014-11-06.

Index

Note: Page numbers followed by the letter *f* indicate material found in figures.